EARTHQUAKE Resistant Building Design and Construction

EARTHQUAKE Resistant Building Design and Construction

Second Edition

Norman B. Green, M.S.
Structural Engineer

Van Nostrand Reinhold Company Regional Offices:
New York Cincinnati Atlanta Dallas San Francisco

Van Nostrand Reinhold Company International Offices:
London Toronto Melbourne

Library of Congress Catalog Card Number: 80-25507
ISBN: 0-442-28799-2

Manufactured in the United States of America

Published by Van Nostrand Reinhold Company
135 West 50th Street, New York, N.Y. 10020

Published simultaneously in Canada by Van Nostrand Reinhold Ltd.

15 14 13 12 11 10 9 8 7 6 5 4 3 2

Library of Congress Cataloging in Publication Data

Green, Norman B.
Earthquake resistant building design and construction.

Includes bibliographical references and index.
1. Earthquakes and building. 2. Earthquake resistant design. I. Title.
TH1095.G73 1981 693.8'52 80-25507
ISBN 0-442-28799-2

Preface to First Edition

It is the purpose of this book to give a general overall view of earthquake engineering as it is applied to buildings, and to present all pertinent information concerning earthquakes. The latest earthquake-resistant design concepts are treated, including ductility, damping, and the influence of energy absorption by structures.

Earthquake-resistant design is evolutionary, and, although great progress has been made since seismic design was made mandatory by building codes, it is still not completely understood. These unknowns, including what is probably the most important—that is, the relation of static to dynamic forces—are discussed herein.

The book is written from the standpoint of the practicing structural engineer, and a special effort has been made to present and explain practical methods of analysis and design that can readily be used in a design office. The book will also be of interest to the student of engineering.

Acknowledgment is due to the California Institute of Technology and the American Iron and Steel Institute for photographs and to the Portland Cement Association for permission to use parts of their publication, *Shear Wall-Frame Interaction a Design Aid.* Professor H. Bolton Seed has very kindly permitted the use of an Acceleration Diagram and an Acceleration Response Spectrum that appeared in one of his papers.

NORMAN B. GREEN

Preface to Second Edition

Changes have been made throughout this edition and much new material has been added.

The general subject of gunite and guniting has been amplified because it is not well covered in most of the engineering literature. An important feature of this book is the presentation of a practical method for the design of the combination of shear walls and frames and the application of this method to the design of a bracing system is described in detail in this edition. There is added information on earthquakes including terminology, magnitude, fault activity, and, particularly on liquefaction, which, it is now recognized, can cause extensive damage within a city.

There is a description of very recent cyclic loading tests of epoxy-repaired concrete beams. These tests are important, because nearly all such testing has previously been with static loads, whereas cyclic loads are necessary to simulate the actual effects of an earthquake.

In compiling this edition, part of the new material that has been added includes methods of analysis and descriptions of designs, so that the book is of even greater interest to the practicing structural engineer.

NORMAN B. GREEN

Contents

EARTHQUAKE Resistant Building Design and Construction

1

Earthquakes: Their Importance to the Engineer

In recent years, structural engineers have been giving more and more attention to the design of buildings for earthquake resistance. A great deal of time and effort has gone into the development of better methods of design. This has involved a greater knowledge of earthquakes in conjunction with a better understanding of the forces they exert on buildings. New concepts have been developed concerning the earthquake resistance of buildings as determined by their ability to absorb the energy input from the earth vibration.

THE ORIGIN OF AN EARTHQUAKE

An earthquake originates on a plane of weakness or a fracture in the earth's crust, termed a "fault." The earth on one side of the fault slides or slips horizontally and/or vertically with respect to the earth on the opposite side, and this generates a vibration that is transmitted outward in all directions. This vibration constitutes the earthquake.

The earthquake generally originates deep within the earth at a point on the fault where the stress that produces the slip is a maximum. This point is called the hypocenter or focus and the point on the earth's surface directly above this point is called the epicenter. The main or greatest shock is usually followed by numerous smaller aftershocks. These aftershocks are produced by slippage at other points on the fault.

TYPES OF EARTHQUAKE FAULTS

Faults are classified in accordance with the direction and nature of the relative displacement of the earth at the fault plane. Probably the most common type is the *strike—slip* fault in which the relative fault displacement is mainly horizontal across an essentially vertical fault plane. The great San Andreas fault in California is of this type. Another type is termed a *normal fault*—when the relative movement is in an upward and downward direction on a nearly vertical fault plane. The great Alaskan earthquake of 1964 was apparently of this type. A less common type is the *thrust fault*—when the earth is under compressive stress across the fault and the slippage is in an upward and downward direction along an inclined fault plane. The San Fernando earthquake was generated on what has usually been classified as a thrust fault, although there was about as much lateral slippage as up and down slippage due to thrust across the inclined fault plane. Some authorities refer to this combined action as lateral thrust faulting. The compressive strain in the earth of the San Fernando Valley floor just south of the thrust fault was evidenced in many places by buckled sidewalks and asphalt paving.[1]

FORCES EXERTED BY AN EARTHQUAKE

Slippage along the fault occurs suddenly. It is a release of stress that has gradually built-up in the rocks of the earth's crust. Although the vibrational movement of the earth during an earthquake is in all directions, the horizontal components are of chief importance to the structural engineer.

These movements exert forces on a structure because they accelerate. This acceleration is simply a change in the velocity of the earth movement. Since the ground motion in an earthquake is vibratory, the acceleration and force that it exerts on a structure reverses in direction periodically, at short intervals of time.

The structural engineer is interested in the force exerted on a body by the movement of the earth. This may be determined from Newton's second law of motion which may be stated in the following form:

$$F = Ma$$

In which F is a force that produces an acceleration a when acting on a body of mass M. This equation is nondimensional. For calculations M is set equal to W/g, then:

$$F = W/g\ a \tag{1}$$

In which F is in pounds, a is in feet per second per second, W is the weight of the body also in pounds and g is the acceleration of gravity, which is 32.2 feet per second per second.

Equation (1) is empirical. It simply states the experimental fact that for a free falling body the acceleration a is equal to g and the acceleration force F is then equal to the weight W.

For convenience, the acceleration of an earthquake is generally expressed as a ratio to the acceleration of gravity. This ratio is called a seismic coefficient. The advantage of this system is that the force exerted on a body by acceleration is simply the corresponding seismic coefficient multiplied by the weight of the body. This is in accordance with Equation (1) in which a/g is the seismic coefficient. For example, if the acceleration of an earthquake is 3.22 feet per second per second, the seismic coefficient is 0.10. If a body weighing 100 pounds rests on the earth, the seismic force exerted on it by the earthquake is 100 times 0.10 equals 10 pounds. During an earthquake the ground acceleration varies continually in an irregular manner. Accordingly, for each earthquake there is a maximum acceleration.

LENGTHS OF FAULTS AND EXTENT OF EARTH MOVEMENT

There are numerous earthquake faults in California and particularly in Southern California. The lengths of these faults vary from about 20 miles to as much as 600 miles, which is the approximate length of the great San Andreas fault. The lengths of faults are difficult to determine since the slippages that produce earthquakes generally occur deep in the earth and the disruption along the fault often does not reach the surface. Among different earthquakes there is a wide variation in the length of the fault on which slippage occurs. In the San Fernando earthquake it has been estimated that this length was about 10 miles, whereas during the 1906 San Francisco earthquake, the slippage was along a length of fault about two hundred miles long. Different earthquakes also show a wide variation in the amount of slippage on the fault. During the San Fernando earthquake, the maximum slippage across the fault was five to six feet and there was about the same amount of horizontal slippage along the fault. Large earthquakes have involved a much greater movement or displacement on the fault. For example, during the 1906 San Francisco earthquake, there was a surface movement or offset of 21 feet at a point on the fault just north of San Francisco.

The permanent slippage of the earth on a fault should not be confused with the displacement of the earth due to vibratory movement. Although the slippage along the fault may be several feet as it was in the San Fernando earthquake, the vibratory movement is generally only a few inches. For example, in the 1940 El Centro earthquake the maximum displacement due to vibration was about six inches. This vibratory displacement from an earthquake is not permanent.

After the severe shaking, the earth returns to its original position.

ACTIVITY OF FAULTS

All faults are not considered to present the same hazard. Some are classified as "active" since it is believed that these faults may undergo movement from time to time in the immediate geologic future. Unfortunately in the present state-of-the-art there is a good deal of uncertainty in the identification of potentially active faults. For example, the fault that generated the San Fernando earthquake did not even appear on any published geological maps of the area. This fault was discovered to be active only when it actually slipped and ruptured the ground surface. Accordingly the identification of active faults and geologically hazardous areas for land use criteria and for hazard reduction by special engineering may be of questionable value.

Only in very recent years have geologists begun to try to evaluate the potential activity of faults that have no historical record of activity. By close inspection of a fault, visible in the side walls of a trench that cuts across the fault, it is sometimes possible to determine if it has been active in recent times. For example, if the trace of the fault extends through a recent alluvial material, then there must have been slippage since that material was deposited. However fault ruptures may be very difficult or impossible to see in unbedded material such as sand and gravel. Also of course the location of the fault must be known and it must reach the surface of the ground in order to inspect it by trenching.

Evidence of the historical activity of a fault may sometimes be obtained by observing the faulting of geologically young deposits exposed in a trench. Such deposits are generally bedded and well consolidated so that fault rupture can easily be seen.

The approximate time of formation of a fault rupture or

scarp has in some cases been determined by radiocarbon analysis of pieces of wood found in the rupture or scarp.

In addition to evidence of young fault activity obtained by trenching, there also may be topographic evidence of young faulting such as is obvious along the San Andreas fault. Vertical aerial photographs are one of the most important methods for finding topographic evidence of active faults. This evidence, which includes scarps, offset channels, depressions, and elongated ridges and valleys, is produced by fault activity. The age of these topographic features and therefore the time of the fault activity, can be estimated by the extent to which they are weathered and eroded.

TECTONICS

Recently the theory of tectonics has been used to explain the underlying mechanics at work in generating earthquakes. According to this theory the earth's outer shell consists of huge plates up to 60 miles in thickness which float on a partially plastic layer of the upper mantle. These plates are presumed to move laterally and grind together at their margins thus producing earthquake faults. For example, on the west coast of the United States the Pacific plate is involved in a grinding collision with the North American plate. This has produced the great San Andreas fault. The earth's crust west of the San Andreas fault is slowly creeping northward with respect to the earth on the east side of the fault, at the rate of about 1¾ inches per year.

The San Fernando earthquake can be explained by tectonic theory.[19] It resulted from the movement of the San Gabriel mountains plate of the earth's crust north of the fault, in a southwestward direction. This exerted a compression and shear against the San Fernando Valley plate south of the fault. The mountain plate was uplifted as much as seven feet and shifted westward as much as six feet relative to the valley plate. This movement of a plate of the earth's crust produced the San Fernando fault and a zone of tectonic ruptures at the ground surface along the north margin of the San Fernando Valley.

The theory of tectonics neatly explains earthquakes along existing plate boundaries. These occur on the earthquake prone ring that circles the Pacific Ocean. It does not explain midplate quakes far distant from plate margins, such as the very severe earthquake that occurred in 1811 and 1812 in the Mississippi Valley, centered near New Madrid in southeastern Missouri. This quake is believed to have been of a Richter magnitude of about 8.0 and at the epicenter the maximum Modified Mercalli intensity verged on the maximum on the scale, or about XII.

An interesting result of research on midplate earthquakes east of the Rocky Mountains is that the eastern quakes have a much larger "felt" area than quakes in the far west. For example, the New Madrid quake was felt as a Modified Mercalli intensity of *V* or more over an area of 2.5 million square kilometers, whereas the 1906 San Francisco earthquake gave a minimum intensity of *V* over an area of only about 150,000 square kilometers. Thus, although the eastern quakes are relatively infrequent, they may be damaging over a much greater area. The reason for this difference is not known.

Scientific understanding of seismic activity west of the Rocky Mountains is far greater than it is in the east. In fact no well accepted mechanism, such as tectonics, that produces the eastern quakes has been proposed. Of course it is due to slippage on faults, but the exact location and extent of the fault system in the New Madrid area is not known. On the other hand, many of the active faults in the west are known.

THE RICHTER MAGNITUDE

The magnitude of an earthquake is commonly designated by a number on the Richter scale. The Richter Magnitude is a measure of the maximum amplitude of the ground vibration as recorded by a seismograph. There is no direct relationship between the energy release and the Richter number. However in general, as the magnitude increases so does the energy of the earthquake, which is related to the length of fault on which slippage occurs. The Richter scale of magnitudes is logarithmic, so that an increase of unity on the scale, represents a ten fold increase of magnitude. For example, a reading of 6.0 is ten times greater than a magnitude 5.0 quake. A 7.0 reading is 100 times the magnitude of a 5.0. It should be recognized that this Richter number is not a good measure of the damage potential of an earthquake. For example, the San Fernando earthquake had a Richter magnitude of 6.6, but the ground acceleration in the region of greatest shaking was as great and the damage to structures was probably nearly as great as would be produced by an earthquake of magnitude of 8.0. The difference is that the severe ground shaking would be of a longer duration for the greater earthquake and it would extend over a much greater area. The longer duration of shaking would probably produce somewhat greater damage to structures.

MODIFIED MERCALLI INTENSITY

The Modified Mercalli Intensity Scale originated by Woods and Neumann in 1931 (see Appendix) is widely used in the United States and in some other countries as a measure of the damage potential of an earthquake. Ratings on this intensity scale are based on peoples reactions to earth movement, on the observed damage to structures and on the observation of other physical effects. Since these observations involve both judgment and dependability and because the design and construction of structures varies throughout the world and even in different parts of the United States, intensity ratings under this system may not be very consistent. A scale of intensities has been proposed which bases the numbered ratings on actual measured ground accelerations. These ratings should be reasonably consistent and should give information of direct use to the structural engineer. The trouble is that no measured accelerations are available for many earthquakes.

EARTHQUAKE ACCELERATIONS AND STRONG MOTION ACCELEROGRAPHS

Much more rapid progress has been made in the development of methods for the earthquake resistant design of buildings, since the development and use of the strong motion accelerograph. This instrument can provide a continuous record of the earthquake acceleration of the ground or of a building, in the region of greatest shaking. This is called an accelerogram (see Figure 8). This is exactly the information that is of use to the structural engineer.

Until shortly before the San Fernando earthquake, there were few strong motion accelerograph installations in California and only two of these accelerographs had been subjected to an earthquake of significantly high acceleration. The first of these two earthquakes, which was in 1940 near El Centro, California, had a magnitude of 7.1 and a maximum acceleration of 0.33g. The second earthquake, which was in 1966 at Parkfield, California, had a maximum acceleration of 0.50g, but the magnitude was only 5.5 and the duration of severe shaking was very short.

The acceleration at the Parkfield earthquake was the highest ever recorded, the previous highest having been at El Centro. This very high acceleration came as somewhat of a shock to structural engineers. Heretofore, practically all investigations of the actual earthquake forces acting on a structure and the manner of its vibration and deflection, had been based on the record of the 1940 El Centro earthquake. Considering the high acceleration in the Parkfield earthquake, it was easy to predict that much higher accelerations than the 0.33g at El Centro, were likely to occur in future earthquakes of large magnitude. The merit of this prediction was soon demonstrated by the San Fernando earthquake.

From the standpoint of the structural engineer, the San Fernando earthquake is the most important that has ever occurred. Far more strong motion data has come from this earthquake than from any previous one. This data was provided from the acceleration records of 272 strong motion accelerographs that had been recently installed in southern California. Over half of these accelerographs were installed in 60 recently constructed multistory buildings, in accordance with a new building code requirement adopted by the city of Los Angeles in 1965. Under this code requirement three accelerographs were installed in each of these buildings, one on the first floor or in the basement if there was a basement, one at midheight and one on the roof. These accelerograph records in buildings are particularly important. They give not only the acceleration at ground level, but they also show how the building vibrated and deflected. Very few such records in buildings had been obtained before. Previously the studies of building vibrations were based on those produced by mechanical vibrators which generate extremely small deflections.

The maximum recorded acceleration of the San Fernando earthquake was 1.25g at the Pacoima dam, which is about three miles south of the epicenter of the main shock. The next closest record was at the seven story Holiday Inn, approximately 14 miles south of the epicenter. At the ground floor of this building the maximum acceleraton was 0.27g and it was 0.40g at the roof. There were no accelerographs in the northern part of the San Fernando Valley, but an estimate can be made of the peak ground accelerations in this area from the records mentioned. On this basis it has been estimated that the peak acceleration in the northern part of the valley was "in the approximate range of 0.30 to 0.50g."[1] In accordance with calculations made by the Author, this estimate appears to be somewhat low. These calculations give an acceleration of 0.83g from the breakage of bracing rods in a building in the San Fernando Valley Juvenile Facility, which is about 7.5 miles south of the epicenter of the main shock.[2]

Although measured earthquake ground accelerations are very important, particularly as a basis for dynamic analyses of structures, it must be recognized that they do not have any direct correlation with the seismic coefficients that are used in engineering design. The most important factor in determining the values of these coefficients is observation by competent engineers, of the actual damage to structures. These values are only a fraction of the peak ground accelerations that strong motion instruments would measure during an earthquake.

LIQUEFACTION

Earthquake vibration of saturated fine granular soil, sands and silts, may produce what is termed *liquefaction*, which is a physical process whereby the material develops fluid consistency. Structures resting on the liquefied soil will have partial or complete loss of support of their foundations and may also sustain damage due to subsidence of the surface of the earth due to soil compaction. The remedial measures to prevent liquefaction are costly. These include remove and replace, grouting, compaction by vibration or dewatering.

In June 1964, there occurred an earthquake at Niigata, Japan, which is an example of the tremendous damage that can result from soil failure due to liquefaction, if it occurs over a wide area within a city. This city is situated on the loose sandy soil of a river plain, where there is a high water table so that the soil is saturated at all times. Although this earthquake was of a magnitude 7, the acceleration was only 0.15 g at Niigata which was not enough to do extensive structural damage to well built buildings such as existed in this city. The damage, nearly a billion dollars, was almost entirely a result of foundation failure due to liquefaction which caused buildings to settle and tip over, as shown by Figure 1A.

The San Juan, Argentina, earthquake of 1977 having a magnitude of 7.4, is of particular interest with regard to liquefaction. The most important effect of this earthquake was the liquefaction of the soil in low lying areas having a high water table and covering several hundred square kilometers. Unlike the liquefaction at Niigata, however, there were relatively few buildings in these areas. The structures

Figure 1A. Buildings in Niigata, Japan, overturned due to foundation failure from liquefaction during the 1960 earthquake.

in the liquefied areas were damaged by subsidence and particularly by differential subsidence of as much as 12 to 14 inches.

In July 1976, an earthquake measuring 7.8 on the Richter scale occurred near the city of Tangshan in China. The members of a Chinese delegation to a recent engineering conference at Stanford University described for the first time the appalling devastation and loss of life caused by this major earthquake in a heavily populated city not built to resist it. Chinese engineers reported extensive failures of building and bridge foundations due to liquefaction of the soil beneath them. They stated that the problem of soil liquefaction in an earthquake is unresolved. No measures have been found to prevent damage from it.

It is estimated that the violent shaking causing water saturated soil to liquefy over an area of several thousand square miles. This liquefaction was accompanied by the usual "sand boils," that is, spouts of a mixture of sand and water erupting from the ground. An American engineer who visited China after the earthquake stated that these sand boils were so numerous, that the sand was spread thickly over large areas devoted to agriculture, which ruined the harvest. This is an interesting type of damage caused by liquefaction, which is generally considered to be damaging only to structures.

OTHER EARTHQUAKE EFFECTS

Although vibration and acceleration are the most important earthquake effects as far as the structural engineer is concerned, there are other effects that must be considered. These are disruption of the surface of the ground due to faulting, strain in the ground at the surface, liquefaction of foundation soils, and land slides. The San Fernando earthquake displayed all of these effects.

There are two reasons why vibration and acceleration are much more important to the structural engineer than are the other effects described. In the first place, vibration and acceleration are a part of every earthquake, whereas the other effects do not always occur. Also the structural engineer can do little or nothing to protect a structure from damage due to the other effects, he can only recommend that structures not be built where they are likely to occur. For this purpose the structural engineer should have the advice of professionals in other fields, namely soils engineering, geology and seismology.

No structure should be built on a fault, whether or not the fault is believed to be active. If the disruption of the surface of the ground due to faulting occurs under a building it will be seriously damaged. Nothing can be done to prevent this.

Often strains in the earth occur adjacent to a fault that ruptures the surface of the ground during an earthquake. The strains may be due to shear stress or to tension and compression that produce changes in length. Such compressive changes occurred during the San Fernando earthquake as previously mentioned. These effects caused broken curbs, broken pipes, and buckled pavements as well as building damage.[1]

The term land slide as used here has nothing to do with angle of repose. It means the movement of earth often on a very flat slope, due to sliding over an underlying unstable material. This instability may be due to liquefaction or simply to the nature of the underlying material. In Anchorage Alaska, during the 1964 earthquake, there were very extensive land slides due to an underlying unstable clay. The most dramatic earthquake destruction in Anchorage was due to these slides which did enormous structural damage. In one slide area alone, more than seventy of the finest homes in the city were destroyed.

In an area severely shaken by the San Fernando earthquake a land slide was caused by liquefaction. The slide area was about 900-feet wide and a mile long and the horizontal movement ranged from a few inches to about 3 feet. The surface on which this movement occurred had a very gentle slope, only about 1.5 degrees. Only a few buildings were affected by the slide but it crossed a main highway and a railroad line which were severely damaged.

TSUNAMI

If the epicenter of an earthquake is below the ocean it may produce seismic waves called tsunami. Such waves will not be produced if the earth displacement is primarily lateral and not vertical. For example, a severe earthquake having a Richter magnitude of 7.7 occurred in Peru in May 1970. The epicenter of this earthquake was 25 miles off shore, but there was no significant tsunami associated with this earthquake. On the other hand, in November 1975, there was an earthquake of magnitude 7.2 on the island of Hawaii. The epicenter of this earthquake was right on the shoreline, where a wave was generated seconds after the earthquake, having a height of 20 feet above the normal water level. The height of this first wave diminished rapidly along the shoreline, roughly in proportion to the distance from the epicenter, so that on the far side of the island it was only about 3 feet.

The first indication of a tsunami is generally a severe recession of the water, which is shortly followed by a returning rush of water that floods inland a distance depending on the height of the wave. This recession and return of the water continues at intervals as each wave of the usual series arrives at the beach. These tsunami are long period waves that may travel great distances from the point of generation.

The most destructive tsunami in Hawaii occurred on April 1, 1946, following an Aleutian Islands earthquake. Waves of 55 feet in height, crest to trough, struck the northeast coast of Hawaii at Hilo, 173 persons were killed, 488 buildings were demolished and 936 more were damaged. Damage was estimated at 25 million dollars.

A tsunami from this same Aleutian earthquake caused considerable changes in the water level of Los Angeles harbor. These changes produced strong currents in a narrow channel of water behind Terminal Island in the harbor. Some damage was done due to the movement of small Naval craft that were moored to a wharf in this channel. The Author was aware of this at the time because he was in the Navy, stationed at the Naval Base on Terminal Island.

The 1964 Alaska earthquake was accompanied by a major tsunami that swept from the Gulf of Alaska, across the length and breadth of the Pacific Ocean. This tsunami was generated mainly by the uplift of a very large area of the continental shelf. It caused extensive damage to shore installations in Alaska and British Columbia and took 16 lives in Oregon and California.

EARTHQUAKE PREDICTION

A good deal of research on earthquake prediction is being carried out in the United States, in Japan, in the Soviet Union, and in the People's Republic of China. The present status of this work is described in a report entitled "Earthquake Prediction and Public Policy," published in 1975 by the National Academy of Sciences, which is quoted as follows:

> Earthquake prediction is still in a research stage: networks of instrumentation needed for dependable prediction are inadequate except perhaps in one small area of California, and historical baseline data against which to detect the premonitory signs of a large earthquake are not yet available. But theory and experience are advanced enough to

justify confidence that an expanding prediction capability may be imminent. By prediction we mean the specification of place, time and magnitude of earthquakes within sufficiently narrow limits to permit short-term and long-term actions to save life and property.

A less optimistic opinion concerning earthquake prediction was recently expressed by the eminent seismologist Charles F. Richter. In the March 1976 issue of the Newsletter Earthquake Engineering Research Institute, Dr. Richter made the following statement:

> Until a few years ago one could say that no clear progress had been made in the directon of prediction. Then work in the Soviet Union indicated changes in the speed of seismic waves, presumably related to the accumulation of strains in the rock. The Russians and others have founded optimistic hopes of prediction on this and other types of observation. However, the present state of knowledge does not warrant hopes of exact and useful prediction within a year, or five years, or twenty years.

The "premonitory signs" mentioned above are movements and physical changes in the earth's crust that may precede an earthquake. These signs include the rate of creep of two fault plates past each other, the amount of tilt of the earth, the amount of strain accumulating in the earth's crust as revealed by precise surveys, changes in elevation, changes in the speed of seismic waves, changes in fluid pressure, and electrical conductivity in rocks and changes in the frequency of small earthquakes. All of these possible premonitory signs are under investigation.

Effective prediction, particularly with regard to place or location, requires a knowledge of all the active faults where strong earthquakes may originate, but this information is not yet available. The San Fernando earthquake was on a fault which had been mapped only imperfectly and which was not generally regarded as a hazard. When the Oroville earthquake occurred in 1975, it was not known what fault produced the shock. The location of the fault zone responsible for the earthquake was only revealed by a series of after shocks and by a zone of surface fracturing that occurred during the earthquake. In the past there have been other earthquakes that could not be assigned to known faults.

Concerning the maximum magnitude of the earthquake that will occur on any given fault, it may be expected that it will be at least as great as the maximum earthquake that has occurred on the fault in the past. A determination of the probable maximum magnitude therefore, requires a relatively long history of earthquakes of varying magnitudes that have occurred on the fault and for which the magnitudes have been recorded. The number of such recorded earthquakes that is required for a reliable statistical determination of maximum magnitude is not generally available.

The prediction of the probable time of occurrence of an earthquake is subject to even more uncertainty than the other two parameters. In the first place, this prediction must be entirely dependent on premonitory signs and these signs must be recorded on an extensive network of instruments. As mentioned above, existing instrument networks are inadequate. Also, since an earthquake might occur immediately after a change in some sign, the observation and recording of signs must be continuous. The establishment of any useful correlation between the recorded premonitory signs and the time of occurrence of earthquakes, will certainly require many earthquakes and long periods of observation.

The movements that occur in the earth's crust at a fault and which are measured and studied as possible premonitory signs, are illustrated by changes in elevation and creep that have occurred at the San Andreas fault in California.

A "bulge" has gradually developed between 1959 and 1974, along a 200 mile stretch of the San Andreas fault, centering near Palmdale, California. The maximum uplift is about 18 inches and the total area covered by the bulge is about 32,000 square miles. There has been some recent depression of parts of the bulge. For example, at Palmdale, the maximum rise since 1959 of about 14 inches, has reduced to 7 inches since 1974. These episodes of uplift and collapse present a geological puzzle the significance of which has not been determined. While land uplift is known to have preceded destructive earthquakes as in Niigata, Japan, in 1964, there has been no record of such movement before other major earthquakes and such uplift has occurred without being followed by any earthquake. The uplift is still continuing and is being closely observed by the U.S. Geological Survey.

Geodetic surveys show that the earth's crust west of the San Andreas fault is slowly moving or creeping northward with respect to the earth on the east side of the fault. This relative movement or creep is a maximum south of Hollister at about 1¾ inches per year. This sounds very small, but it represents about 7 feet over a 50 year period. The creep diminishes southward until it is nearly unmeasurable in the vicinity of Los Angeles. The fact that the rate of fault movement or creep often changes prior to small earthquakes, encourages the belief that with increased frequency of measurement, such changes may ultimately provide a means for predicting the occurrence of destructive earthquakes.

GEOLOGIC AND SEISMIC EVALUATION OF A SITE

For an important building project it is now common practice for the owner to employ consultants in the fields of geology, seismology and soils engineering, to prepare a geologic-seismic report on the building site. This is for the purpose of providing the structural engineer with information concerning the geologic nature of the site and the potential for earthquake damage to the building or buildings that will be constructed.

Such reports are expected to give the locations of active faults up to 100 miles from the site, that could affect the site and also estimates of the location and magnitude of earthquakes that may occur on those faults. It should be recognized that such information is actually earthquake prediction and it is subject to all the limitations and uncertainties that pertain to such predictions.

2

History and Development of Seismic Building Codes

California is a highly seismic region. California engineers must continually deal with this condition; consequently, they are always in the forefront in developing methods for the design of earthquake resistant structures. Therefore professionals from this geographic area have more or less dominated the field of earthquake resistant design. Engineers in other areas look to the California engineer for what is being done and what is being recommended. Accordingly the story of the development of seismic building codes in general, may very well be confined to a description of their development in California.

The earthquakes that will be referred to as having had a major influence on the development of seismic codes are all California earthquakes. This is because they have been experienced and their effects have been observed by many residents of California, including consulting engineers and state, county and municipal officials, who are directly concerned with building codes and damage to buildings.*

*This chapter is hopefully a better approach to the subject of earthquake resistant design, than to simply describe the latest code. This step by step approach will clarify the many problems as they developed and should make it easier to understand the end result, which is a rather sophisticated code.

EARLY DEVELOPMENTS

Probably the greatest earthquake that we have had in California, at least in a heavily populated area, occurred in 1906 in San Francisco. Strangely enough, this earthquake did not result in any changes in California building code requirements with regard to lateral forces. Even after this event, buildings were still designed to resist wind, if there was any lateral force design, and no consideration was given to earthquake forces. In fact, in Los Angeles the building code did not have any requirement for either wind or earthquakes until 1924, when the code was revised to require design for wind pressure. This situation may perhaps be attributed to the erroneous assumption that if a building is designed for an adequate wind force, it will also have adequate earthquake resistance.

In 1925 a severe earthquake hit Santa Barbara, California. This gave an impetus to earthquake studies and research in the United States. Structural engineers in California began to give serious attention to earthquake resistant design and papers on the subject began to appear in engineering journals. The California building codes, however, still did not include earthquake design requirements to a significant extent. It was not until the Long Beach, California, earthquake in 1933 that mandatory seismic codes were first published in the United States.

The Long Beach earthquake was severe and it did extensive damage in Long Beach and to a lesser extent in the environs, including Los Angeles. There was an immediate and very strong reaction by the California Legislature, by municipal officials, and by practicing structural engineers and architects. Everyone was shocked by the damage done in Long Beach and particularly by the damage done to school buildings. The only thing that prevented a heavy loss of life among school children was the fact that the earthquake occurred at 6:00 A.M., when the schools were not in session.

In 1933 the State Legislature passed the Riley Act, which required that all buildings, with a few minor exceptions, must be designed to resist a relatively small earthquake force. The Field Act was also passed in 1933, which gave the State Division of Architecture jurisdiction over the structural design of all school buildings and the rules and regulations were adopted to govern this design. These rules and regulations were much more severe in their requirements than the Riley Act and they constituted one of the first really comprehensive codes for earthquake resistant design. This code also contained rigid requirements for the testing of materials for school buildings and for the field inspection of their construction. Also in 1933 the counties and the municipalities in the State revised their building codes to include earthquake design requirements.

The first building code provisions for earthquake resistant building design specified a single seismic coefficient for determining the design forces. This coefficient was the same regardless of the type of construction or height of the building. Generally higher coefficients were specified for certain individual parts of the bulding, such as parapet walls, chimneys, tanks and exterior ornamentation. Experience had shown that these portions of a building are particularly susceptible to earthquake damage.

THE FIRST LOS ANGELES CITY SEISMIC CODE EARLY JAPANESE REQUIREMENTS

In 1933 the first earthquake provisions were incorporated in the building code of the City of Los Angeles. These provisions were of the type described above, with a basic seismic coefficient of 0.08, which was increased to 0.10 for schools.

In 1932 Dr. Kyoji Suyehiro gave a series of lectures in this country in which he stated that buildings in Japan were designed using a seismic coefficient of 0.10.[3] He also stated that buildings designed in this way resisted the very severe Kwanto earthquake in 1923 "fairly well." Evidently the Japanese had seismic requirements in their building codes some ten years before they appeared in our California codes.

This method of calculating the earthquake forces acting on a building may be expressed as follows:

$$F = CW \tag{2}$$

In which F is the earthquake force acting on any portion of a building, W is the weight of that same portion and C is the seismic coefficient.

THE EFFECTS OF BUILDING DEFLECTION

The use of a single seismic coefficient for calculating the earthquake forces acting on a building is correct only if the building and its foundation are rigid, so that all parts of the building move with the ground. The entire building is then subjected to the same acceleration and it is proper to use a single seismic coefficient for its design.

Actually buildings are not rigid. During an earthquake a building deflects and vibrates, with the deflection and acceleration progressively increasing from the first floor to a maximum at the roof. Also the more flexible a building, that is, the longer its natural period of vibration, the smaller are the accelerations produced in it by an earthquake. This is practically the same as saying the higher the building the smaller are the seismic accelerations. This relationship of period to acceleration response is readily apparent by inspection of the acceleration response spectrums (see Figure 9).

INFLUENCE OF BUILDING HEIGHT ON SEISMIC FORCES THE 1943 LOS ANGELES CITY CODE

The next step in the development of the earthquake provisions of building codes, was to make the equivalent static design loading more in line with the actual dynamic loading that produces the response described above. This was done by using a variable seismic coefficient C, instead of the single fixed value that had previously been used. An example of this change is the building code that was adopted by the City of Los Angeles in 1943. In this code the seismic coefficient was expressed as follows:

$$C = \frac{60}{(N + 4.5)100} \tag{3}$$

In this equation, N is the number of stories above the one under consideration. Considering the top story of a building or a one story building, N would be zero. The coefficient C was used in the following equation:

$$V = CW \tag{4}$$

In which V is the shear in the story under consideration and W is the total weight of the N stories of the building above the story under consideration.

BASE SHEAR

From Equation 4 it is seen that the shear V is simply the total earthquake force acting above the story under consideration. From Equations 3 and 4 the shear V can be determined for a one story building or for each story of a multistory building. A very important shear in a multistory building is what is termed the "base shear," which is the shear in the first story. This is the total earthquake force acting on the building.

In accordance with Equations 3 and 4, the seismic acceleration and the forces acting on a multistory building increase progressively from the first floor to a maximum at the roof. Thus the higher the building the smaller will be the seismic coefficient for the base shear. For example, the base shear coefficient is 0.045 for a ten story building and the corresponding seismic acceleration is 0.045g. For a twenty story building the corresponding values are smaller, namely 0.027 and 0.027g. In other words, the static earthquake design forces calculated from Equations 3 and 4, will produce a building response that approximates the actual earthquake dynamic effects more closely than does the response produced by the design forces prescribed by previous codes. That is to say, more closely than forces calculated from Equation 2.

REVISION OF THE LOS ANGELES CITY BUILDING CODE FOR BUILDINGS OVER 13 STORIES

A further improvement in this respect, for higher buildings, was attained by a new seismic loading equation that was incorporated in the Los Angeles City building code in 1957. This revision was occasioned by the removal of the 13 story height limit for buildings in Los Angeles. It was thought that Equation 3 could be improved for buildings over 13 stories. This new equation was recommended to the Los Angeles Board of Building and Safety, by a committee of structural engineers appointed in 1956 by the president of the Structural Engineers Association of southern California. The Author was a member of this committee. The new Equation was as follows:

$$C = \frac{4.6\,S}{[N + 0.9(S - 8)]100} \qquad (5)$$

In which C and N are as previously defined and S is the total number of stories in the building, except S equals 13 for buildings of less than 13 stories.

Equation 5 is identical to Equation 3 for buildings of 13 stories or less. The committee thought this was a desirable result, since the new equation would not render buildings obsolete that had been designed under the old Equation 3. For buildings over 13 stories, and particularly for much higher buildings in the 20 to 30 story range, Equation 5 gives a loading that much more closely approximates loads in accordance with dynamic theory. This met one of the objectives of the committee, which was to develop an equation that would give more realistic loading than did Equation 3 for buildings over 150 feet or 13 stories, which had previously been the height limit for buildings in the City of Los Angeles.

By this time the consensus was that the design earthquake loading on a building should be what has been termed "triangular." That is, the forces acting on a uniform multistory building should vary in magnitude uniformly from zero at the first floor to a maximum at the roof. A uniform building is one without set-backs. Strictly speaking, it is the seismic acceleration that should vary in this manner. Equation 5 approximates this type of loading fairly well, except for the lower buildings.

THE SEAOC SEISMIC CODE

In 1957 a committee of the Structural Engineers Association of California was formed under the chairmanship of William T. Wheeler, to develop a new seismic code, the

SEAOC code. Preparation of the code required about two years of work by the committee.*

An important objective of the new code was to develop earthquake resistant design of buildings based on static forces, which was the method used in all previous codes, so that it would be brought in closer agreement with the essential features of dynamic theory. It was also hoped to develop a uniform seismic code which would replace the several different codes that were used in the United States and particularly in California.

Ductility

This new code introduced "ductility," a concept that had not previously appeared in seismic building codes. *High ductility* is the ability of a building to sustain large deflections without failure or collapse. This is a very important characteristic of a building since it greatly reduces the effect or "response" that is produced in the structure by an earthquake. This is because the building is set in vibration by the energy of an earthquake and this vibration and the accompanying deflection is reduced by the energy that is absorbed by the large inelastic deflections of a ductile structure. Methods of earthquake resistant building design now recognize that in a very severe earthquake a building structure will be stressed beyond the so-called elastic limit. The *elastic limit* may be defined as the limit beyond which the structure will sustain a permanent deflecton. For a single member this may be a clearly defined limit, as it is for a structural steel member. For a structure containing many members, however, it is not so clearly defined, because all the members do not reach their elastic limits at the same time.

*The Author was a member of this committee and did a good deal of work on it during those two years.

The measure of ductility is the *ductility factor*, which is defined as the ratio between the maximum deflection without failure and the yield deflection of a structural system.

Base Shear Coefficient

Different types of building construction have varying degrees of ductility, and this is taken care of in the equation for base shear, by introducing a factor K. The equation then becomes:

$$V = KCW \qquad (6)$$

In this formula V is the base shear, W is the total weight of the building above the first floor and C is the base shear seismic coefficient, which is given by the following equation:

$$C = \frac{0.05}{\sqrt[3]{T}} \qquad (7)$$

In this equation T is the period of vibration of the building. To be more exact, it is what is known as the "fundamental" or longest period.

In Equation 7, C is established as a function of the period of vibration of the building: the shorter the period or stiffer the structure, the larger the coefficient and the greater the forces produced in the building by an earthquake. This is in accordance with dynamic theory. In the loading equations (Equations 3 and 5) that appeared in previous codes, there is no direct relationship between vibration period and seismic coefficient. The older equations simply made the coefficient vary with the number of stories in the building. This established an indirect and inexact relationship between coefficient and period of vibration, since a higher building generally but not always, has a longer period of vibration.

Period of Vibration

The application of this new method for calculating the earthquake forces acting on a building required a determination of its vibration period. Any reasonably exact determination of the period of a multistory building involves a complicated and time consuming calculation. Also this calculation cannot be made until the building is designed, but the building cannot be designed until the period is known. Accordingly, for practical use, a simple but inexact equation was devised: Equation 8. It was thought that errors in period due to the use of this equation would not seriously affect the calculated value of the base shear.

$$T = \frac{0.05\,H}{\sqrt{D}} \tag{8}$$

Here H is the height of the building in feet and D is the dimension of the building in feet in a direction parallel to the applied forces. For the purpose of computing C, T need not be less than 0.10. This equation represents a curve that gives a fair average of the plotted values of the measured periods of vibration of several hundred buildings.[4] The measurements were made by the U.S. Coast and Geodetic Survey.

Distribution of Base Shear

For design purposes it was necessary to distribute the base shear into equivalent static forces to be concentrated at each story. As a building vibrates and deflects during an earthquake, the acceleration at each story is directly proportional to the deflection at that level if the deflection does not exceed the elastic limit. The seismic force at any story is then equal to the acceleration at that story times the mass of the story. This is in accordance with dynamic theory. Therefore, a distribution of the base shear that is in accordance with dynamic theory can be made if the deflection of the building is known. If the deflection is not known, a distribution can be made based on an assumed deflection. This was done by the committee. The committee thought that even if the building structure was stressed beyond its elastic limit, the lateral load distribution would not be seriously affected.

It was assumed that the deflection of a building during an earthquake varies as a straight line from zero at the first floor to a maximum at the roof. The committee believed that this was a good approximation for most buildings. If the weight of each story is the same, this results in a triangular load distribution. The following equation pertains to this method of load distribution:

$$F_x = \frac{V\, w_x h_x}{\sum wh} \tag{9}$$

In which F_x is the lateral force applied to a level designated by x, V is the base shear, w_x is the weight at the level designated by x and h_x is the height in feet above the base, to level x. The Σwh is the summation of the products of all $w_x h_x$'s for the building.

Overturning

Equation 9 is based on the assumption that all the forces F_x act in the same direction, which would be the case if the building vibrated exclusively in its fundamental mode. The overturning moment M at any level is then equal to the total moment exerted by all the forces F_x acting above that level. To take account of the influence of the higher modes of vibration for which the lateral forces are not all in the same direction, see Figure 10, Chapter 4, the committee decided to multiply the moment M by a factor J. This factor would reduce M for buildings having relatively long periods of vibration, that it was believed would be subject to significant vibrations in the higher modes.

Ductility Coefficients

The committee established values for K for four different types of construction, as follows:

$K = 1.33$ is a structural system without a complete vertical load carrying space frame. In this system, the required lateral forces are resisted by shear walls or braced frames. A braced frame is a truss system or its equivalent which is provided to resist lateral forces and in which the members are subjected primarily to axial stresses.

$K = 1.00$ All building framing systems except as classified for $K = 1.33$, $K = 0.80$ or $K = 0.67$.

$K = 0.80$ Buildings with a dual bracing system consisting of a ductile moment-resisting space frame and shear walls or braced frames designed in accordance with the following criteria:

1. The frame and shear walls or braced frames shall resist the total lateral force in accordance with their relative rigidities considering the interaction of the shear walls and frames.
2. The shear walls or braced frames acting independently of the ductile moment-resisting space frame shall resist the total required lateral force.
3. The ductile moment-resisting space frame shall have the capacity to resist not less than 25% of the required lateral force.

$K = 0.67$ Buildings with a ductile moment-resisting space frame having the capacity to resist the total required lateral force.

The four different types of building structure are described above in the order of their increasing resistance to seismic forces. The Box Type structure has the lowest resistance and requires the largest base shear seismic coefficient. These K values are largely based on the actual observed performance of buildings in earthquakes. Types of construction which have "performed" well in the past were assigned lower values of K and, conversely, structures which have not performed well and appeared to be inherently weak in resisting earthquake forces, were assigned higher values of K. These K values are really judgment factors, based on the collective opinions of the members of the committee. As already mentioned, the K values probably reflect rather closely the relative ductilities of the different types of construction.

No way has been developed to measure and specify the ductility of a building structure as it actually performs. A moment resisting space frame is simply required to be made of ductile material or a ductile combination of materials. The ductility of a single structural member can be measured; it is the ratio of total deflection to yield deflection. In order to provide some clarification of the concept of ductility as applied to structures, the following statement appeared in the code. "The necessary ductility shall be considered to be provided by a steel frame with moment resisting connections, or by other systems proven by test and studies to provide equivalent energy absorption."

Ductility of Concrete Frames

This statement resulted in a good deal of controversy, since it was believed that it would practically rule out the use of moment resisting concrete frames for the construction of earthquake resistant buildings. Concrete itself is not a ductile material. The Portland Cement Association immediately went into action. They did a great deal of testing and spent a lot of money in an effort to prove that if concrete is reinforced properly, the combination of concrete and reinforcing steel may be considered as ductile. The upshot was that additions to the SEAOC code were made which described in detail the reinforcing that was required for concrete members in order to render them ductile. However, there is still doubt in the minds of some engineers, including the Author, whether or not this change complies with the original code requirement. In other words, does a concrete frame, even though it is reinforced in accordance with the code requirements, provide "equivalent energy absorption?"

Since the SEAOC code was first published in 1959, a number of changes have been made in it by subsequent seismology committees of the SEAOC. However, the basic criteria and the general approach have not been changed. The most important change was the inclusion in the code of the requirements for a ductile moment resisting concrete frame.

Revision of Overturning Requirement

Another very important change that was made in 1969, was the increase of the design overturning moments by elimination of the *J* factor. This change was made following the publication of reports concerning the compressive failures of columns in several multistory buildings in Caracas, Venezuela, during the 1967 earthquake.[5] These buildings were reinforced concrete, ranging in height between 10 and 21 stories, and with frames that supported the vertical loads and also resisted the lateral forces. It was apparent that the column failures were mainly due to overturning, since they were generally in outside wall columns forming the ends of bracing bents and usually in relatively high and narrow buildings. These buildings were well constructed of good materials, including 4000 pounds per square inch of concrete in the columns, and these failures indicate much higher overturning moments than would have been used for a seismic design under the UBC Code, Zone 3, or the SEAOC Code. See an analysis of one of these failures in Chapter 5.

AD HOC COMMITTEE

In May 1970 the Board of Directors of the SEAOC established an "Ad Hoc Committee" to review the basic seismic design criteria of the SEAOC Code. Since the code was first published in 1959, several important earthquakes have occurred which led to the thought that they should be studied to see if new information might indicate the need for code changes. The chairman of this Ad Hoc Committee was H. J. Degenkolb, consulting structural engineer of San Francisco. The other members were members of the original 1957 committee that prepared the SEAOC Code and the members of the 1970 Seismology Committee of the SEAOC. As a member of the original committee, the Author was also a member of the new Ad Hoc Committee. The Ad Hoc Committee made its report in October 1971, about eight months after the San Fernando earthquake. This earthquake greatly stimulated the work of the committee, since it made it apparent that important changes in the code would have to be made.

During the deliberations of the Ad Hoc Committee, extending over more than a year, each member submitted to the Chairman written comments, opinions and discussion, on each of fifteen questions concerning the basic seismic design criteria of the code. The recommendations, contained in the committee report prepared by the Chairman, were intended to express the consensus of the committee concerning these questions. The twenty-five members of the committee were engineers well versed in the theory and practice of the earthquake resistant design of buildings.

Basic Recommendations

1. It was almost unanimously agreed that the present form of the code, whereby "equivalent static forces" are stipulated based on a general dynamic analysis of a uniform structure, is the most practical form of the code.
2. It was recommended that dynamic analyses be required for major structures and dynamically nonuniform structures, to seek out and reinforce weak or vulnerable places, because adequate criteria are not available.
3. The majority voted for increased design forces.
4. If ductility is not assured, then major force increases are necessary.
5. The committee recommended that the present K factor be retained in the code, but it should be reevaluated.
6. Generally commented upon, is the necessity for considering the effect of the underlying soils. It was strongly agreed that this should be considered in any future code.
7. The function of a building or structure must be recognized in the level of performance specified by the code, using public safety criteria. These structures would include hospitals, centers of emergency, communications and control, and evacuation facilities.

To summarize, the structural effects of the San Fernando earthquake made it clear that changes in the SEAOC Code would be necessary. Several modern multistory concrete buildings located in areas of very strong shaking which were designed in accordance with the seismic requirements of the code, sustained very severe structural damage. For example, the six story Indian Hills Medical Center (see Figure 15) has required extensive and costly repairs (about 16 percent loss) and the seven story Holy Cross Hospital (see Figure 17) required partial demolition (about 50 percent loss). At the Olive View Hospital, the first story of a large two story building completely collapsed and the very large seven story main building was so severely damaged it required demolition.[2]

The peak accelerations recorded in some buildings were well above code requirements, but as mentioned these accelerations do not appear to have any direct correlation with the structural damage to the buildings. For example, the seven story Holiday Inn had a peak roof acceleration of 0.40g whereas the SEAOC Code provision is equivalent to a roof acceleration of only 0.15g. The structural damage was not as great as the expected damage from this acceleration. It simply consisted of some cracking of the reinforced concrete frame members, which was repaired by the use of an epoxy adhesive. However, this deformation was associated with violent movement of the building which caused extensive and costly nonstructural damage. This anomalous result with regard to structural damage may at least partially be explained by the fact that the 0.40g acceleration was very brief, being recorded simply by a single spike or peak in the acceleration diagram.

REVISION OF THE SEAOC CODE

The 1974 edition of the SEAOC Code, entitled Recommended Lateral Force Requirements and Commentary, was published in May 1975. In this new edition, the fourth, major revisions have been incorporated in general accordance with the new recommendations of the Ad Hoc Committee.

The new equation for base shear is:

$$V = ZIKCSW \tag{10}$$

It is seen that three new terms have been added to the original equation, Equation 6.

Zoning Factor

The term Z is a zoning factor, which is intended to take into account that geographic areas generally differ from each other with regard to the likelihood of earthquakes and also with regard to their probable frequency and intensity, if earthquakes have occurred in the past. The seismicity of an area, for zoning purposes, is determined primarily by the historical record of earthquakes and the location, length, and estimated activity of earthquake faults in the region. If there is little or no likelihood of an earthquake the value of Z might be zero. The original SEAOC Code was intended to apply to California conditions of seismicity, for which a single zone was considered to be appropriate and the Z factor was assumed to be unity. The 1974 edition of the code is intended to apply to areas of "highest seismicity" for which the value of Z is again set at unity. It is expected that the revised code will be used more widely, therefore, zoning must be considered, since some areas where it will be used may not have highest seismicity. The Z factor would then be less than unity.

Ordinarily any seismic code that employs a zoning factor, such as the Uniform Code, is used in conjunction with a map showing the geographical extent of each zone of seismic intensity. The SEAOC Seismology Committee did not arrive at any conclusions on geographic limits of Z coefficients. However, studies are continuing in an effort to define specific Z values for various geographic areas.

Increment for Essential Facilities

The term I establishes higher seismic design factors for facilities deemed essential to public welfare and which should remain functional for use after a major earthquake. This is in agreement with recommendation number 7 of the Ad Hoc Committee Report. These essential facilities would include hospitals, fire stations, and communication and power stations. The San Fernando earthquake clearly demonstrated that such facilities should have special consideration. Four hospitals in the San Fernando area were so severely damaged that they were no longer operational, just when they could have been needed the most. The 110 million dollar Sylmar Converter Station was completely disabled, resulting in a 22 million dollar loss and the outage of this very important electrical facility for over a year.

The code establishes a maximum value for I of 1.5 for essential facilities and a value of 1.0 for other structures. No values of I are established for specific facilities. Studies on this coefficient are being continued by the Committee.

Site Factor

The term S is a soil or site factor. Introduction of this term in the base shear equation is in accordance with recommendation number 6 of the Ad Hoc Committee Report. From accelerograph records it is known that there may be significant differences in the ground motion at sites a relatively short distance apart. This has generally been attributed to differing conditions of the underlying soil. However, recent studies indicate that a determination of the so-called soil effect, or site effect, is a complex problem. It involves not only the soil conditions at the site but also the source mechanism of the earthquake, the distance from source to site, surface and subsurface topography and the travel path of the earth vibrations from source to site. The determination of the value of S in the SEAOC Code is based on an estimate of the effect of the first of these influences, which is the effect of soil conditions at the site.

The influence of the soil condition at the site is presumed to be directly related to the ratio of the fundamental vibration period T of the structure on the site, to an inherent natural vibration period T_s of the site itself. An equation is given which makes S a function of this ratio. If the ratio is 1.0 the value of S is 1.5. This represents resonance between structure and site and gives the maximum value of S. The minimum value of S is stated to be unity.

At the present time, all authorities do not agree that every site has an inherent natural vibration period. For example, Professor Paul C. Jennings, California Institute of Technology, recently made a comprehensive study of this matter from accelerograph records of the San Fernando earthquake and also from the records of several earthquakes that have occurred at El Centro, California.[6] The following quotation is from Professor Jennings' report: "Furthermore, the significance of such concepts as the fundamental period of the soil are questionable for the sites studied because they cannot be confirmed by measurements of surface motions during strong earthquakes. It is obvious that such concepts should be applied with caution at sites where there are no recorded accelerograms."

Increase In Seismic Coefficient

Probably the most important change that has been incorporated in the new SEAOC Code is an increase in the value of the coefficient C. This is in compliance with recommendation number 3 of the Ad Hoc Committee report. The new equation for C is as follows:

$$C = \frac{1}{15\sqrt{T}} \tag{11}$$

This gives the greatest increase in earthquake design forces for short period or more rigid buildings. The increase in C

from Equation 11 over the original Equation 7, is 60 percent for a T of 0.30 seconds and 13 percent for a T value of 2.50 seconds. The maximum value of C is set at 0.12.

Ductility Factor

The factor K has been retained in the base shear equation, as recommended by the Ad Hoc Committee. However, some changes have been made in the description of the structural systems required for certain K values. These changes and the latest descriptions are shown on page 32.

The principal change is the inclusion of braced frames and a description of such a frame. A braced frame is similar to a shear wall with regard to stiffness and ductility. It is an important structural system. As mentioned in Chapter 6, the only practicable way to strengthen a multistory steel frame is by adding braces, and the use of K or X bracing for this purpose makes it a braced frame. The only other important change is a much more detailed description of the $K = 0.80$ structural system and how it should function.

Dynamic Analysis

The Ad Hoc Committee recommended that a dynamic analysis be required for major structures. The new requirement in the code is as follows. "The distribution of the lateral forces in structures which have highly irregular shapes, large differences in lateral resistance or stiffness between adjacent stories or other unusual features shall be determined considering the dynamic characteristics of the structure." This appears to be a better requirement than simply a flat statement that a dynamic analysis must be made for all major structures. In practice, under the new code requirement, a large majority of buildings will be designed by the conventional method of equivalent static forces.

Unfortunately highly irregular structures or those having unusual features, for which the Code requires dynamic analysis, are exactly the structures for which it is difficult or impossible to make a good realistic dynamic analysis.

COMPREHENSIVE SEISMIC DESIGN PROVISIONS OF THE APPLIED TECHNOLOGY COUNCIL

In October 1971 an organization entitled, "The Applied Technology Council" (ATC) was formed and incorporated by authorization of the SEAOC. This is a nonprofit corporation for the purpose of transforming the results of research into practice.

The SEAOC made a major contribution to the science and practice of earthquake engineering by development and updating of the "Recommended Lateral Force Requirements and Commentary," which is referred to above as the SEAOC Seismic Code. In recent years there has been a rapid development of new information on seismic design, and it is no longer practicable to assimilate and incorporate such new material in the SEAOC Seismic Code by voluntary activity of committees of the SEAOC. This situation was brought to a head by the very large amount of new information that has developed from the San Fernando earthquake. Accordingly, in December 1973 a project ATC-3, entitled "Comprehensive Seismic Design Provisions," was initiated by the Applied Technology Council.[15]

Production of the SEAOC Seismic Code was almost exclusively the work of practicing structural engineers. In fact, the only departure in this respect was the production of the Site Factor *S*, which was included in the 1974 edition of the code. This involved other professions, soils engineering and seismology.

The work on project ATC–3 is divided among five task groups, as follows.

Task Group 1—Seismic input
  Risk assessment
  Ground motion and site effects
Task Group 2—Structural
  Structural design, details
  Structural analysis
  Soil structure interaction
Task Group 3—Nonstructural
  Architectural systems
  Mechanical, electrical systems and equipment
Task Group 4—Liaison and Format
Task Group 5—Existing buildings
  Inspection and evaluation of damage
  Inspection and evaluation of hazard
  Repair and strengthening

The practicing structural engineer will be chiefly concerned with the work of task groups 1 and 2, which will produce the design recommendations of the new code. Engineers will also be greatly interested in information that may result from the work of task group 5, relative to the repair and strengthening of existing buildings.

In contrast with the production of the SEAOC Seismic Code, there are 41 members in task groups 1 and 2, of which only five are practicing structural engineers who must represent the people that will actually use the code. The other people are necessary for their special knowledge concerning seismology, geology, and soils engineering. However, this make up of the personnel could result in an impracticable code, unless the engineers are able to insist on simplicity. The state-of-the-art in these other fields is not as well developed as it is in the engineering field, that is to say with reference to ground motion and site effects.

Since the Applied Technology Council was formed in 1971 and the project ATC-3 was initiated in 1973, the document ATC-3-06 as described hereinafter, was completed and published in June 1978. This document can be obtained from Roland L. Sharpe, Executive Director, Applied Technology Council.[22]

ATC-3-06 TENTATIVE PROVISIONS FOR THE DEVELOPMENT OF SEISMIC REGULATIONS FOR BUILDINGS

This massive document comprised of 505 pages, was produced by the concerted effort of a multidisciplinary team of nationally recognized experts in earthquake engineering. As indicated by the title, the Applied Technology Council consider the provisions of this document to be tentative in nature and it is recommended by the Council that their viability should be established by trial designs for representative types of buildings prior to their use for regulatory purposes.

This document ATC-3-06 is not by any means a building code even though it is here included in Chapter 2. It contains much material that would not be incorporated in a building code but which is of interest to any engineer concerned with the design of earthquake resistant buildings and with the problem presented by hazardous existing buildings in high earthquake risk areas.

Seismic Performance Categories

An important feature of ATC-3-06 is the assignment of buildings to several different categories in accordance with concern for life safety. The category to which a building is assigned, together with its configuration, determines primarily what method of analysis should be used for designing its seismic resisting system. These Seismic Performance Categories are as follows:

A. Good quality of construction materials and adequate ties and wall anchorage. Limited to low earthquake risk areas. Only wind design is required.
B. Some earthquake resistant requirements but these are quite simplified as compared to requirements in areas of high seismicity.
C. The earthquake resistant requirements compare roughly to present design practice in California seismic areas for buildings other than schools and hospitals.
D. This construction is required for critical structures in relatively high seismic zones. As much redundancy as possible should be provided for the bracing systems of buildings in both categories C and D.

Increment for Essential Facilities

It will be noted that in Equation 20 there is no term like *I* in Equation 10 of the SEAOC Seismic Code, which provides an increment of base shear for essential facilities. In ATC-3-06 it is stated that the improved performance characteristics desired for more critical occupancies are provided by the design and detailing requirements for each Seismic Performance Category and by more stringent drift limitations. It is believed that increasing the force alone does not necessarily increase the performance.

Building Configuration

Past earthquakes have repeatedly shown that buildings having irregular configurations suffer greater damage than buildings having regular configuration. There is plan configuration and vertical configuration. Plan irregularities are, L, T, U, and H shapes. Vertical irregularities are primarily due to set-backs. A building having a regular plan configuration could be square, rectangular or circular.

Drift Limitation

In this document the purpose of drift limitation is to control nonstructural damage for life safety not for economic reasons. The permissible drift varies with the Seismic Performance Category and is a result of the relationship between the use of the building and the level of shaking to which it may be exposed. The allowable story drift is either 1.0% or 1.5% of the story height. The story drift is simply the difference between the displacement of the floor of the story and that of the floor next above. The drift is determined by calculating the deflection of the seismic resisting system by elastic analysis, using the prescribed seismic forces and considering the building to be fixed at the base, and then multiplying the result by the deflection amplification factor C_d as given in Figure B.

Orthogonal Effects

A very unusual requirement in ATC-3-06 is that structural elements be designed for 100% of the effects of seismic forces in one principle direction combined with 30% of the effects of seismic forces in the other principle direction. This requirement is based on the fact that earthquake forces act in both principle directions of a building simultaneously but the effects in the two directions are unlikely to reach their maximums at the same time. Orthogonal effects are slight for beams and girders but may be significant for columns.

Strength of Members and Connections

The document ATC-3-06 presents methods of analysis for determining the forces acting on a building structure. The method of design that is used for the members and connections that resist these forces, acting in combination with other prescribed loads, must be compatible with the total force level.

For reinforced concrete, design is by ultimate strength and ACI 318-71 is referenced. For structural steel, use is made of maximum strength design and the reference is to AISC—Part 2. For wood, working stress method of design is employed using design stresses set at 200% of basic working stresses and the UBC—Chapter 25 and NDS (National Design Specification) are referenced. For masonry, working stress procedures are also used in which the stress is the conventional working stress modified to be comparable to the yield point of more ductile materials. No standard specification for masonry is available and the requirements are stated very extensively in Chapters 12 and 12A.

These design procedures are in accordance with present practice, which recognizes that an earthquake may stress a building structure beyond its elastic limit thereby mobilizing its ductility.

Analysis by Equivalent Static Forces

This method of analysis is recommended in ATC-3-06 as a minimum for regular and irregular buildings assigned to category B, also as a minimum for buildings classified as regular and assigned to categories C and D.

The base shear V is calculated from the following equation:

$$V = C_s W \tag{19}$$

where C_s is the seismic coefficient and W is the weight of the building above the first floor, including 25% of the live load for storage or a warehouse.

The seismic coefficient is as follows:

$$C_s = \frac{1.2\, A_v S}{R T^{2/3}} \tag{20}$$

The value of C_s need not exceed 2.5 A_a/R. For type S_3 soil when A_a is equal to or greater than 0.30 the value of C_s need not exceed 2 A_a/R where

A_v = coefficient representing effective peak velocity-related acceleration (EPV)
S = coefficient for soil profile characteristics of the site.
R = Response Modification factor
T = fundamental period of the building
A_a = coefficient representing effective peak acceleration (EPA)

Large maps are provided giving values of A_a and A_v by different colored areas. These maps are for the United States, Alaska, Hawaii, and Puerto Rico. Each color has a number corresponding to values of A_a and A_v, expressed in decimal fractions of g the acceleration of gravity. These values are as follows:

Area Number	A_a	A_v
7	0.40	0.4
6	0.30	0.3
5	0.20	0.2
4	0.15	0.15
3	0.10	0.10
2	0.05	0.05
1	0.05	0.05

The southern half of California up to Madera County including coastal areas up to Humboldt County is No. 7. Other portions of California are less seismic.

The term S in Equation 20 is a soil or site factor. It has the value 1.0 for S_1 soil, 1.2 for S_2 soil and 1.5 for S_3 soil. The soil characteristics are as follows:

S_1 = Rock of any characteristic, either shale-like or crystaline in nature, or stiff soil conditions where the soil depth is less than 200 feet and the soil types overlying rock are stable deposits of sand, gravel or stiff clays.

S_2 = Deep cohesionless or stiff clay soil conditions, including sites where the soil depth exceeds 200 feet and the soil types overlying rock are stable deposits of sands, gravels or stiff clays.

S_3 = Soft-to-medium stiff clays and sands, characterized by 30 feet or more of soft-to-medium stiff clay with or without intervening layers of sand or other cohesionless soils.

The *R* term in Equation 20 serves the same purpose as the factor *K* in SEAOC Seismic Code, although *R* is intended to account for both ductility and damping inherent in the structural system. Values of *R* are given in Figure B. The ratio between the maximum and minimum values of *K* is only 2 whereas this ratio is nearly 8 for *R*. Some explanation should be presented for this wide difference, since the values of both *K* and *R* represent the consensus of engineers skilled in earthquake engineering.

The fundamental period *T* is calculated by use of the established methods of mechanics assuming the base of the building to be fixed. An approximate period T_a is given by one of the following formulas:

$$T_a = C_T H^{3/4} \tag{21}$$

Formula 21 applies to moment-resisting structures where the frames are not enclosed or adjoined by more rigid components tending to prevent the frames from deflecting when subjected to seismic forces.

C_T = 0.035 for steel frames
C_T = 0.025 for concrete frames
H = The height of the building in feet

Type of Structural System	Vertical Seismic Resisting System	Coefficients R	C_d
Bearing Wall System: A structural system with bearing walls providing support for all, or major portions of, the vertical loads. Seismic force resistance is provided by shear walls or braced frames.	Light framed walls with shear panels	6½	4
	Shear walls		
	Reinforced concrete	4½	4
	Reinforced masonry	3½	3
	Braced frames	4	3½
	Unreinforced and partially reinforced masonry shear walls[6]	1¼	1¼
Building Frame System: A structural system with an essentially complete Space Frame providing support for vertical loads. Seismic force resistance is provided by shear walls or braced frames.	Light framed walls with Shear panels	7	4½
	Shear walls		
	Reinforced concrete	5½	5
	Reinforced masonry	4½	4
	Braced frames	5	4½
	Unreinforced and partially reinforced masonry shear walls[6]	1½	1½
Moment Resisting Frame System: A structural system with an essentially complete Space Frame providing support for vertical loads. Seismic force resistance is provided by Ordinary or Special Moment Frames capable of resisting the total prescribed forces.	Special moment frames		
	Steel[3]	8	5½
	Reinforced concrete[4]	7	6
	Ordinary moment frames		
	Steel[2]	4½	4
	Reinforced concrete[5]	2	2
Dual System: A structural system with an essentially complete Space Frame providing support for vertical loads. A Special Moment Frame shall be provided which shall be capable of resisting at least 25 percent of the prescribed seismic forces. The total seismic force resistance is provided by the combination of the Special Moment Frame and shear walls or braced frames in proportion to their relative rigidities.	Shear walls		
	Reinforced concrete	8	6½
	Reinforced masonry	6½	5½
	Wood sheathed shear panels	8	5
	Braced frames	6	5
Inverted Pendulum Structures. Structures where the framing resisting the total prescribed seismic forces acts essentially as isolated cantilevers and provides support for vertical load.	Special Moment Frames		
	Structural steel[3]	2½	2½
	Reinforced concrete[4]	2½	2½
	Ordinary Moment Frames		
	Structural steel[2]	¼	1½

Figure B. Response modification coefficients[1]

For all other buildings

$$T_a = \frac{0.05\ H}{\sqrt{D}} \tag{22}$$

which is the same as Formula 8 of the SEAOC Seismic Code.

It is stated that T_a should be used to calculate an initial base shear from which a preliminary design can be made. It is evidently intended that the final design be based on the calculated value T. Engineers are accustomed to use the approximate value of the period for the final design.

The vertical distribution of the lateral forces is either linear or parabolic, depending on the period of the building. The linear distribution is used in the SEAOC Seismic Code in accordance with Equation 9, which is recommended in ATC-3-06 only for building periods of 0.50 second or less. For parabolic distribution the same Equation 9 is used except h_x and h each have the exponent 2.0 if the building has a period of 2.5 seconds or more. For periods between 0.50 second and 2.5 seconds the exponent may be taken as 2.0 or it may be determined by linear interpolation between exponents 1.0 and 2.0.

Dynamic Analysis

Two methods of dynamic analysis are presented in ATC-3-06 which are the same as those described in Chapter 4 under the heading Analysis by Mathematical Model and under the sub heading Multidegree of Freedom.

In the second method of analysis, described in ATC-3-06 as Modal Analysis, the building is modeled as a system of masses lumped at the floor levels, each mass having one degree of freedom, that of lateral displacement in the direction under consideration. The analysis is as described in the book, except the effective weight is used instead of the effective mass and this weight is multiplied by C_s from Equation 20 in which T is the period of vibration of the mode under consideration. This factor C_s is based on so-called Normalized Elastic Response Spectra which have been reduced by the factor R so that the resulting base shear need not be reduced for ductile design as was done in the procedure described in the book. Normalized elastic response spectra are similar to Figure 9 with irregularities smoothed out so as to give conservative approximations to average ordinates.

The design values of the base shear, lateral force and deflection at each level are determined by combining their modal values. The combination is carried out by taking the square root of the sum of the squares of each of the modal values.

The modal analysis may be used for Seismic Performance Categories C and D having vertical irregularities. Vertical irregularities are irregularities of mass and stiffness over the height of the building. Setbacks produce this type of irregularity.

The first method of analysis, called Analysis by Mathematical Model, is used for buildings assigned to Seismic Performance Categories C and D classified as irregular in plan, which are located in areas of high seismicity and whose failure would pose significant hazard to the public.

For buildings irregular in plan the seismic effects involve a coupled combination of translational and torsional motion and the Analysis by Equivalent Static Forces or the Modal Analysis are likely to be inadequate. In such cases a more rigorous procedure is desirable if the building is located in an area with high seismicity and whose failure would pose significant hazard to the public. The more rigorous analysis is inelastic and involves step by step integration of the coupled equations of motion. It should be recognized that the results of such a nonlinear analysis are only as good as the mathematical model chosen to represent the building vibrating at large amplitudes of motion. This is the method described in Chapter 4 under the heading Analysis by Mathematical Model, except this method is for elastic action.

Soil Structure Interaction

Fundamental to the design provisions in the Analysis by Equivalent Static Forces and in the Modal Analysis is the assumption that the motion that is experienced by a structure during an earthquake is the same as the so-called "free field" ground motion, a term that refers to the motion which would occur at the level of the foundation if no structure was present. Strictly speaking, this assumption is true only for structures supported on essentially rigid ground, for structures on soft soil there is an important rocking component in addition to the translational component, which is particularly significant for tall structures. The soil structure interaction also has the important effect that a substantial part of the vibrational energy may be dissipated in the supporting soil, which increases the damping.

The soil structure interaction effect should not be confused with the so-called "site effect" which is reflected in the values of the design coefficients employed in the Analysis by Equivalent Static Forces and the Modal Dynamic Analysis.

Soil structure interaction may increase, decrease or have no effect on the magnitude of the maximum forces induced in the structure. However, for the design of rigidly supported structures as described in the static force and modal methods of analysis mentioned above, soil structure interaction will reduce the design values of the lateral forces from the values for the rigid base condition. Accordingly adjustments recommended to allow for soil structure interaction may be conservatively neglected.

Because of the rocking motion of a building on soft soil the deflection of the structure is increased, which affects the required spacing of structures and also increases the P-Delta effect. Both these effects are generally small.

Procedures are given for calculating the reduced base shear and the increased fundamental period of vibration, which may be considered as supplementary to the procedures given in the equivalent static force method of analysis. Procedures are also given for calculating the reduced modal base shear corresponding to the fundamental mode of vibration, which may be considered as supplementary to the procedures of the modal dynamic analysis. No reduction is made in the shear components contributed by the higher modes of vibration.

AN APPRAISAL OF THE CONCEPT OF EQUIVALENT STATIC FORCES

The seismic design of buildings is largely in accordance with the method of equivalent static forces, the development of which has been described in this chapter. Observations of earthquake damage and recorded accelerations during the San Fernando earthquake, provided a means for evaluating the concept of equivalent static forces. It should be noted that the static forces are referred to as equivalent, not equal, to the dynamic forces. In other words, do static forces produce equivalent structural effects?

It is well known that earthquakes produce forces in building structures that are much greater than the design forces prescribed by code, but there is not a corresponding structural damage. Consider, for example, the response of the Holiday Inn to the San Fernando earthquake. As already mentioned, the ground acceleration was 0.276g, the greatest acceleration measured at any building. The corresponding base shear for that building has been calculated from the actual measured accelerations at midheight and roof.[16] These accelerations are for the first mode response so that all lateral forces are in the same direction (see Figure 10). The mode shape is then a smooth line (nearly triangular) starting with the roof acceleration (.20g) passing through the midheight acceleration at the fourth floor (.12g) and ending with zero acceleration at the ground. The indicated accelerations at each level are then multiplied by the floor mass at each level to give the dynamic forces which correspond to the first mode response. The sum of these forces is the base shear.

This calculated base shear was almost 3.5 times the value used in the design of the building in compliance with the building code. It might be expected that a base shear 3.5 times the design shear would cause the collapse of the building. Actually the structural damage to the moment resisting concrete frame of this building was not excessive and it was rather easily repaired. This same base shear calculation was made for a number of other buildings in an area that was strongly shaken by the earthquake. In every case the calculated shear was much greater than the shear that had been used in the design of the building, as much as 4.6 times as great, but the structural damage was not excessive.

There is no clear understanding of the reason for this lack of correlation between the damage to buildings and the dynamic earthquake forces. It is known that the compressive strength of concrete and the yield stress in structural steel are both somewhat higher under dynamic loads, but the difference does not appear to be great enough to explain the anomaly with regard to earthquake damage. The same is true of other factors, such as the added resistance supplied by nonstructural elements such as walls and partitions. It does not appear that either ductility or damping in the building structure are involved, since the dynamic forces were calculated from the actual measured accelerations in upper portions of the building, and these quantities had already been modified by ductility and damping.

Another factor that should be considered, however, is the fast rate and brief duration of loading due to the dynamic forces of an earthquake. This type of loading probably produces less damage than static forces. When the rate of loading is so fast as to be of the nature of an impact the elongation at fracture of mild steel is much higher than with rates such as are common in testing. Under such fast rates or impulsive loading it is probable, too, that the stress sustained at the yield point is greater than that under lower rates. Various materials also record higher stresses before fracture if the load is applied quickly; the values obtained in slow testing may easily be doubled. It must be remembered that the working unit stresses that are employed in the engineering design of buildings are based on common testing procedures which involve relatively slow rates of loading.

It is apparent that the base shear for the Holiday Inn as calculated on an equivalent static force basis in compliance with the building code, is much more realistic as far as design is concerned, than the base shear that was calculated from the actual dynamic forces.

3

Bracing Systems and Load Distribution

GENERAL REQUIREMENTS

There are certain essential requirements for any earthquake resistant building structure. It is necessary to have adequate vertical bracing elements, either frames or shear walls, that give stability to the structure by transmitting all earthquake forces to the ground. There must also be horizontal elements or diaphragms that tie the structure together and distribute all lateral forces to the vertical bracing elements. A continuous path is necessary for the transmission of each lateral force, from its point of origin to the ground. The point of origin of a seismic force is at the center of gravity of a mass that is being accelerated.

The vertical bracing elements must be supported on foundations that are adequate to resist all downward loads, possible uplifts and horizontal shear forces. It is very important that the individual footings be tied together to prevent any relative horizontal movements. Such movements would change the load distribution among the vertical bracing elements. If there is any horizontal movement the entire foundation should move as a unit.

DISTRIBUTION OF EARTHQUAKE FORCES WITHOUT TORSION

The floor and roof structures of a building must function as the horizontal elements or diaphragms which distribute the lateral earthquake forces to the vertical bracing elements. For buildings having reinforced concrete or steel deck floors and roofs, it is widely assumed that the diaphragms are rigid under the action of the distributing forces acting in the plane of the diaphragm. Actually the assumption is that the in-plane deformations of a diaphragm are not sufficient to significantly affect the load distribution. Therefore, if there is no in-plane rotation of the diaphragms due to torsion, each vertical bracing element will undergo the same lateral deflection at each level. If the bracing elements in a system are all of the same type, either shear walls or frames, the total lateral load will be distributed to these elements in proportion to their rigidities or stiffness factors. The *stiffness factor* of any structural element is the force required to produce unit deflection. If a bracing system includes both shear walls and frames, a somewhat different method of load distribution must be used.

Beam And Column Frames

A common type of vertical bracing element is the frame with moment resisting beam to column connections, which is sometimes referred to as a "rigid frame." The connections are assumed to be rigid, and the lateral deflection of the frame due to shear results from the bending of the beams and columns. The connections of a moment resisting frame of either steel or concrete are actually not rigid, but practical methods for calculating the joint yielding and its effect on the deflection of the frame have not yet been developed. See the sub heading, "Effect of Bar Slip on Joint Yielding," in Chapter 7. There is also lateral deflection of a frame due to axial deformation of the columns and bending of the frame as a whole, see Appendix. However, the shear deflection is predominant, and the bending deflection is often ignored in distributing the lateral load to the frames of a bracing system.

If there is no torsion and there are only frames in the bracing system with no shear walls, the earthquake load may be distributed to the frames in proportion to their stiffness factors. The stiffness factor of a frame, designated by S, is equal to the triangular load that will produce unit deflection at the top of the frame. This triangular load varies from a maximum at the top to zero at the ground, which is the type of load that is employed in the equivalent static force method of seismic design.

In order to derive a practicable method for calculating the lateral deflection of a frame, some simplifying assumptions must be made. The method presented here is based on the following assumptions.

1. The center to center spacing l of columns is constant across the frame.
2. The story height h is constant.
3. Under lateral load alone, the inflection points of the beams are at their centers.
4. The columns have inflection points at midheight.
5. The moment of inertia I_c of the interior columns in a story are all the same.
6. The moment of inertia I_c of an end column in a story is half the moment of inertia of an interior column.
7. The moments of inertia I_b are the same for all the beams at the same level.

If the inflection points are at the centers of the beams, then the beam end slopes are the same across the frame.

These beam end rotations are caused by the column shears times half the story height. An end column rotates one beam and an interior column rotates two beams through the same angle. Since l and I_b are the same for each beam, an end column must develop half the shear of an interior column.

Assumptions numbers 3 and 4 concerning the position of inflection points in the columns and beams of a frame are used in the "Portal" method of analysis for multistory frames subjected to lateral loads. In this method also, an end column is assigned half the shear of an interior column. The portal method has been widely used in the past for the design of large and important buildings, for example, the 50 story Lincoln Building in New York City. The chief error is in the top and bottom stories of a frame, where the inflection points of the columns are not likely to be at midheight. The relatively small effect of this error on the top deflection of a frame may be ignored.

The total deflection of a frame can be determined from the deflections between successive stories. Figure 1 shows the deflected center line of a typical interior column extending between the centers of successive stories of a frame and the elastic lines of the two floor beams framing into the column, all in accordance with the above described assumptions. An end column of a frame would look the same, except there would be a beam framing into only one side of the column instead of a beam on both sides as shown in Figure 1. At the top story, the column would extend only from the lower inflection point to the beam, with no column above the beam.

The first step in calculating the deflection Δ between two successive stories of a frame, is to distribute the total shear V in each story to the columns in the story, so that each interior column gets the same shear and the two end columns each get half the shear of an interior column. This determines the shear forces v_1 and v_2, as shown in Figure 1. The deflection Δ can then be calculated from Equations 12 or 13

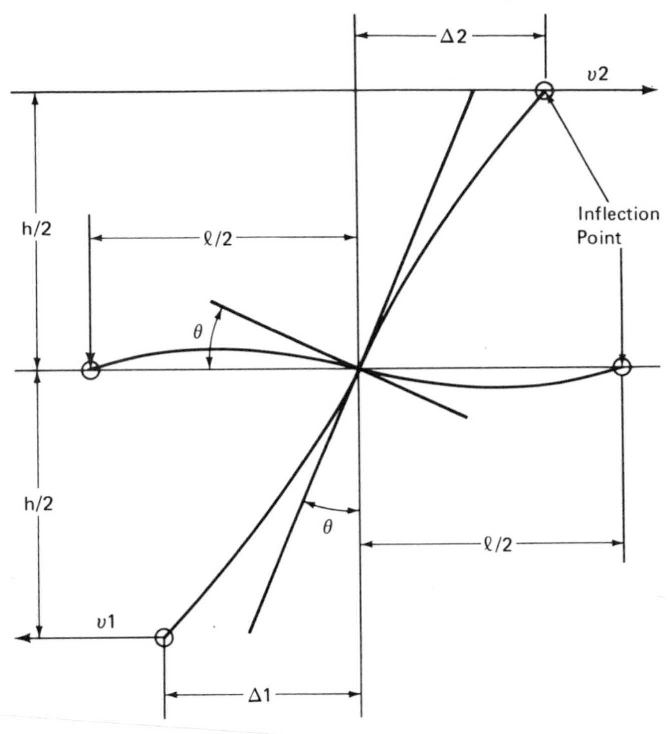

Figure 1. Structural element showing deflection $\Delta_1 + \Delta_2$ between the centers of two successive stories of a frame, due to bending of the column and beam.

and 14. Any column of the frame may be chosen for this purpose, since the deflection Δ will be the same for each of the columns extending between two successive stories.
For an interior column of a frame;

$$\Delta = \frac{1}{24\,E}\left[(v_1 + v_2)\,\frac{h^2 l}{I_b} + (h - a)^3\left(\frac{v_1}{I_{c-1}} + \frac{v_2}{I_{c-2}}\right)\right] \quad (12)$$

For the end column of a frame;

$$\Delta = \frac{1}{12\,E}\left[(v_1 + v_2)\,\frac{h^2 l}{I_b} + \frac{(h - a)^3}{2}\left(\frac{v_1}{I_{c-1}} + \frac{v_2}{I_{c-2}}\right)\right] \quad (13)$$

For the top story end column of a frame.

$$\Delta = \frac{V}{24\,E}\left(\frac{h^2 l}{I_b} + \frac{(h - a)^3}{I_c}\right) \tag{14}$$

The notation in Equations 12, 13 and 14 is as follows;

$\Delta = \Delta_1 + \Delta_2 =$ The deflection between the centers of two successive stories in a multistory frame, or the deflection of the top beam with respect to the inflection points in the top story columns in inches.

$V =$ The shear in any story of a multistory frame in pounds.

$a =$ The depth of the beam. (inches) The effective length of the Column for bending is $h - a/2$.

$v =$ The shear in an end column of the top story in pounds.

$v_1 =$ The shear in the lower half of the column shown in Figure 1.

$v_2 =$ The shear in the upper half of the column shown in Figure 1.

$h =$ Story height in inches.

$l =$ Distance center to center of columns in inches.

$I_c =$ The moment of inertia of an end column in the top story.

$I_{c-1} =$ The moment of inertia of the lower half of the column of Figure 1.

$I_{c-2} =$ The moment of inertia of the upper half of the column of Figure 1.

$I_b =$ The moment of inertia of a beam at the level under consideration, see Figure 1.

$S =$ The stiffness factor of a frame, which is the triangular load on the frame, varying from a maximum at the top to zero at the ground, which produces a unit top deflection of the frame.

E = Modulus of elasticity of structural steel for a steel frame, or the modulus of elasticity of concrete for a concrete frame.

To calculate the stiffness factor S of a frame, the shear in each story due to an arbitrary triangular load is first determined. From these shears the deflection between successive stories can be calculated from Equation 12 or 13 and the deflection in the top story can be calculated from Equation 14. The shear deflection at the top of the frame is then the sum of the deflections between the centers of successive stories, plus the deflection of the top beam relative to the column inflection points in the top story, plus the deflection of the first story columns between their inflection points and the foundation. The triangular load that will produce unit deflection at the top of the frame can then easily be determined by direct proportion.

The calculated deflection of a concrete frame will not be as close to the real deflection as is the case for a structural steel frame within its elastic limit. There are two reasons for this. Strictly speaking concrete does not have a modulus of elasticity since the stress strain line in compression has some curvature even at low stresses. The stress strain line for structural steel in either tension or compression is straight up to the elastic limit. For concrete, the so-called secant modulus of elasticity, which generally is used for stress calculations, should be used. The other reason for less accuracy in the calculated deflection of a concrete frame, is that the moment of inertia of a member that should be used, is uncertain. The effective moment of inertia of a reinforced concrete member actually changes as the member is loaded, due to progressive cracking. At the start, and for relatively low loads, the concrete is uncracked and the entire section is effective. For the calculation of stresses in reinforced concrete, it is customary to assume that the concrete has no tensile strength and the calculation is based on the "cracked section." In Equations 12, 13 and 14, the moment of inertia of a concrete member should be that of the cracked section. Load tests of concrete members show that a calculated deflection based on the cracked section, is quite close to the measured deflection at high loads when there is extensive cracking. But this is the condition that will develop in the members of a concrete frame during a strong earthquake.

When calculating Δ for a steel frame, from Equations 12, 13 and 14, the moments of inertia of the bare steel beams and columns should be used. In accordance with the usual practice, all of the members of the frame will be fireproofed with Spray On, which has little or no effect on the moment of inertia of the member. Concrete floor and roof slabs or concrete fill on steel deck will probably provide composite action of the beams. However, this increases their moments of inertia only for positive moment and this effect is relatively small for the usual slab thicknesses and for wall beams where the floor is on only one side of the beam, which applies to most bracing bents since they are generally in exterior walls.

Shear Walls Without Openings

Another type of vertical bracing element is a shear wall. These are often used, either alone or in combination with frames, to give a stiffer building. A shear wall acts as a vertical cantilever beam which undergoes lateral deflection due to bending and shear. If the wall has no openings or if the openings are small, the bending deflection will predominate and if the relatively small shear deflection is ignored, the stiffness factor of the shear wall is proportional to the moment of inertia of a horizontal section of the wall. This section should include the vertical reinforcing steel in the end columns or chords of the concrete wall. The section area of this steel should be transformed to an equivalent area of concrete in the usual way, by multiplying by (n−1), where n is the ratio of the modulus of elasticity of steel to that of concrete.

The ratio of bending deflection to shear deflection of a shear wall without openings will increase as the ratio of height to width increases. For a ratio of three between the height of a shear wall and its width, the ratio of bending deflection to shear deflection will be about seven for either the cracked or uncracked section. For the cracked section, it is here assumed that the shear is resisted only by the area under flexural compression. This assumption will give the smallest ratio of bending to shear deflection.

If the vertical bracing elements in a building are composed of concrete shear walls without openings, and there is no torsion, then the total lateral load may be distributed to these walls in proportion to their moments of inertia. Usually the moment of inertia of a shear wall will reduce from the first floor upward, primarily due to reductions in wall thickness. For correct load distribution, the percentage of reduction must be the same for each wall in the building, so that the relationship between the moments of inertia of the several walls is the same in each story.

Walls at elevator shafts, stairwells and central cores, are often used as shear walls and these walls generally do not have openings. This is also true of walls at or adjacent to property lines. In modern buildings with air conditioning, windows are sometimes eliminated and solid shear walls are used as architectural features. If shear walls are used for bracing, it is very desirable that they do not have openings. Shear walls with openings present a much more complex problem for the designer.

It is recognized that the deflection of a shear wall, which is primarily due to bending, may be significantly affected by rotation of a foundation that is on compressible soil. An attempt has been made to include this effect in the deflection calculations, since it would affect load distribution.[17] This is particularly true for distribution between shear walls and frames, since frames deflect mainly in shear and would not be affected to the same extent by foundation rotation. The difficulty with trying to calculate foundation rotation, is that foundation soils are not elastic. Under the first loading the stress strain relationship may be nearly linear, but there is little recovery when the load is removed. Also there is very little information concerning the stress strain relationship for soils under dynamic loads, as produced by an earthquake. It would be difficult to carry out soil tests that would give this information. This would have to be done at each building site.

A better approach to the problem of foundation rotation is to use shear wall foundations that would reduce the rotation to a minimum. On a compressible soil, a combined footing should be used that has a large moment of inertia. Piles or caissons could be used extending to rock or a stiff soil, if such is available at a reasonable depth.

Combination of Shear Walls and Frames

If the bracing system without torsion is comprised of both shear walls and frames, the load cannot simply be distributed between them in proportion to their stiffness factors because their modes of deflection are not the same. A shear wall deflects predominantly in bending and shear deflection is predominant in a frame. This difference is indicated by Figure 2. A special method is necessary for load distribution, when a bracing system combines both shear walls and frames.

For a combined system, the beams that connect a shear wall to a moment resisting frame in the same plane, should have hinged end connections, or something approximating a hinge. The interaction between the two structures is thereby

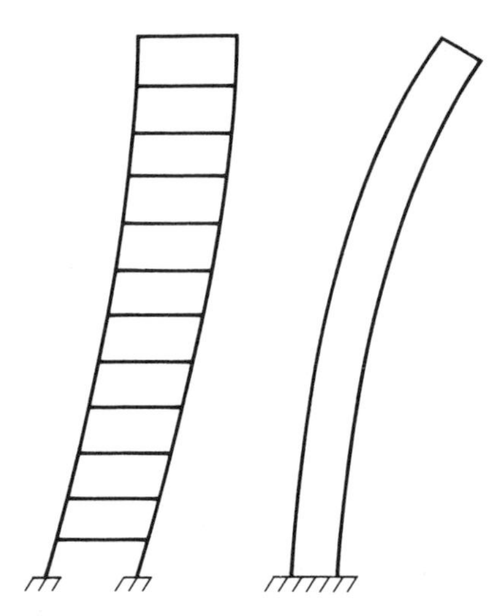

Figure 2. Showing the shear mode of deflection of a rigid frame and the bending mode of deflection of a shear wall.

reduced to a system of horizontal forces. However, these forces are nonuniform and difficult to evaluate. Accordingly a simplified analysis has been devised for use in design, in which the interaction forces acting at each floor are replaced by a single force acting at the top of each frame and shear wall.[17]

The following nomenclature is used in this analysis.

- d = The top deflection of the combined shear wall due to the load W.
- W = Triangular lateral load that is applied to the combined shear wall which varies from zero at the ground to a maximum at the top of the wall. This is the total load acting on the bracing system.
- I_w = Moment of inertia of the horizontal section of a shear wall.
- H = Height of a shear wall in inches.
- K_w = Lateral point load at the top of a shear wall, that produces unit lateral deflection at the top of the wall in pounds.

K_f = Lateral point load at the top of a frame, that produces a unit lateral deflection at the top of the frame in pounds.
E_c = Modulus of elasticity of the concrete of a shear wall.
P = Lateral interacting point load at the top of the combined shear wall and the combined frame in pounds.

The shear walls of the bracing system are replaced by a single combined wall resisting the entire load W and having a moment of inertia equal to the sum of the moments of inertia of the individual walls. The individual frames of the bracing system are replaced by a single combined frame having a value of K_f to be determined in accordance with the desired interaction between the combined wall and the combined frame.

First consider that the combined shear wall is free to deflect independently of the frames, then:

$$d = \frac{11\ WH^3}{60\ E_c I_w} \tag{15}$$

Also, from the definition of K_w

$$\frac{K_w H^3}{3\ E_c I_w} = 1 \text{ then } \frac{H^3}{E_c I_w} = \frac{3}{K_w}$$

and making this substitution in Equation 15

$$d = \frac{11\ W}{20\ K_w} \tag{16}$$

If the combined shear wall and the combined frame acting together under the action of the load W, as indicated in Figure 3, have the same top deflection, then

$$\frac{P}{K_w} + \frac{P}{K_f} = \frac{11\ W}{20\ K_w}$$

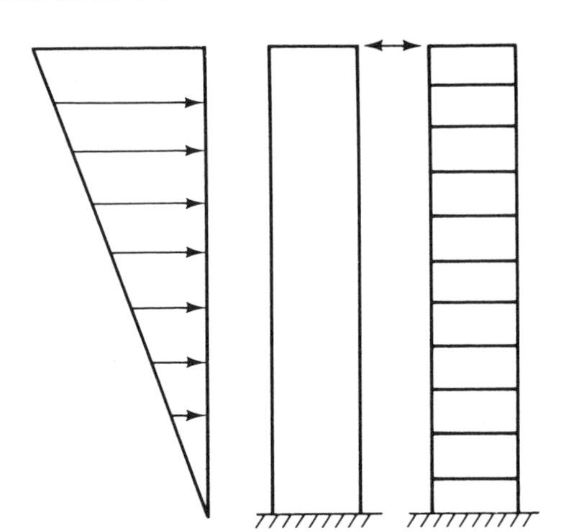

Figure 3. Showing a shear wall subjected to a triangular lateral load and interacting with a frame by a force at the top only.

In this equation K_f is for the combined frame and K_w is for the combined shear wall, then

$$P = \frac{11\ W}{20\left(1 + \dfrac{K_w}{K_f}\right)} \tag{17}$$

To determine the value of K_f for any frame, the top deflection is calculated for any arbitrary lateral point load acting at the top of the frame. This deflection is the sum of the deflections between the centers of successive stories, from Equation 12 or 13; plus the deflection of the top beam from Equation 14; plus the deflection of the inflection points of the first story columns. Having calculated the top deflection of the frame for any assumed top lateral load, the load required to produce a unit top deflection can be determined by direct proportion.

The factor K_w is derived as follows:

$$\frac{K_w H^3}{3\ E_c I_w} = 1 \text{ whence } K_w = \frac{3\ E_c I_w}{H^3}$$

The calculation of K_f may be simplified. Since the shear in the frame is constant and the dead plus live load is probably the same or nearly the same at each floor, the beam size will be the same at each floor, that is, it will be typical. Then the only variation in the size of the members from story to story is the column. Therefore, there will be a relatively small variation in Δ throughout the height of the frame and the total top deflection can be obtained quite closely by multiplying an average Δ by the number of floors in the frame above the first floor. This average Δ can be calculated by using the typical beam and the column size at mid-height of the frame, in the deflection formulas (12 and 13), modified for constant shear. These deflection formulas are as follows:

$$\Delta = \frac{1}{24E}\left(\frac{2\, v_a h^2 l}{I_b} + \frac{2\,(h - a)^3 v_a}{I_{ca}}\right) \tag{12a}$$

$$\Delta = \frac{1}{12E}\left(\frac{2\, v_a h^2 l}{I_b} + \frac{(h - a)^3 v_a}{I_{ca}}\right) \tag{13a}$$

In these formulas v_a is the constant shear, I_b is the moment of inertia of the typical beam, and I_{ca} is the moment of inertia of the column at mid-height of the frame. This calculation of the top deflection does not include Δ in the top story from Equation 14 and the first-story deflection. This simplified calculation gives a close enough value of the top deflection for buildings over four stories. The top load K_f to produce unit top deflection of the frame is then determined by direct proportion.

In order to apply this method of design of the combination of shear walls and frames, it is first necessary to decide how large the force P should be in order to sufficiently reduce the bending moment and shear on the combined wall for good interaction. Reduction of the deflection of the wall is also of some importance.

In accordance with Equation 17, the magnitude of P depends on the ratio K_w/K_f and the smaller this ratio the larger is P. The author of this method of design (Ref. 17) states that it lacks accuracy and should not be used if this ratio is less than 1.0. It is believed that the largest value of P complying with this limitation on K_w/K_f will give the best interaction. Accordingly K_w/K_f should equal 1.0, the smallest allowable value, and then from Equation 17 $P = 0.275\ W$. This value of P will reduce the base moment on the combined wall by 41.5%, will reduce the maximum shear by 27.5% and the top deflection by 50%.

Both the force P and the total lateral load W is distributed to the individual shear walls in proportion to their moments of inertia I_w. The force P is divided between the individual frames in proportion to their K_f's. Two or more frames will be required for each shear wall in order to avoid the necessity for excessively heavy frames. The sum of the K_f's of all the frames must equal the K_f of the combined frame, which is $3\ E_cI_w/H^3$ in which I_w is the moment of inertia of the combined wall. The only easy way to accomplish this distribution is to make all the frames the same and distribute the load P equally among them.

The shear walls are designed for strength to resist these distributed loads. The frames, however, must be designed for both strength and stiffness under the distributed forces P.

When an individual shear wall is re-designed to include the force P the flexural tensile reinforcing will be reduced since the moment is reduced, and this will slightly reduce the I_w (by about 10%) and therefore also the K_w since this is directly proportional to I_w. This reduction of the K_w will also slightly reduce the required K_f for each frame since $K_f = K_w$. These changes may usually be neglected.

When a frame is designed for strength to resist the distributed force P it will be found that it is much too flexible to have the requisite K_f. The members of this frame would therefore have to be very much increased in size and the stresses correspondingly reduced. The resulting frame would be very uneconomical. It may be concluded that frames and reinforced concrete shear walls do not work well together as parts of the same bracing system, because of the difference in intrinsic stiffness between frames and walls.

Comparison with More Accurate Analysis

Calculations have been made of the forces acting between a multistory frame and a concrete shear wall resisting a uniform lateral load, the two structures being forced to have the same lateral deflection throughout their heights due to interconnection at each floor through rigid diaphragms. This interaction is what actually occurs, which is not true of the analysis that equalizes deflection only at the top by means of a single force P. The question is, does this single interacting force P produce sufficiently accurate effects on the shear wall and on the frame to be a useful method of analysis and design.

As mentioned above, the more accurate calculations were for a uniformly distributed lateral load on the shear wall. For this same loading on the wall, the analysis using the single top force P gave results in very good agreement with the more accurate calculations. For example, the bending moment at the base of the shear wall and the top deflection of the frame were practically the same with either method of analysis. The shear on the frame was not in quite as good agreement. The maximum shear by the more accurate analysis, which occurred near mid-height of the frame, was about 30% greater than the constant shear produced by the force P.

Even though these results are not directly applicable to the analysis using triangular earthquake loading, they do indicate that the use of a single top interacting force is acceptable for design purposes.

Since writing the above, the Author wrote a letter to Dr. Iain MacLeod, who developed this method of analysis of a combination of shear walls and frames,[17] suggesting to him that it would be a valuable addition to his paper if the accuracy of the method is determined for triangular earthquake loading, in the same manner as was done for the uniformly distributed load.

Dr. MacLeod responded with a letter from Glasgow, Scotland, where he is a professor of Civil Engineering at Paisley College of Technology. His letter, which stated that he has carried out the accuracy investigation that the Author suggested, also transmitted full information concerning the results. The following quotations from his letter refer to the triangular load investigation.

1. Estimate of frame shear is somewhat less accurate with the triangular load but accuracy is of the same order for both loads.
2. Frame shear is still underestimated in the order of 20 to 40%.
3. There seems to be a general tendency for the error in frame shear to increase with increasing K_w/K_f.
4. Accuracy does not fall off rapidly below $K_w/K_f = 1$ as for uniformly distributed load.
5. Maximum error in the moment at base of wall and the top deflection is of the order of 10%.

The errors are all on the low side. From No. 3 and No. 4 it is apparent that the use of $K_w/K_f = 1$ is correct, since larger values of this ratio would give a larger error in frame shear and accuracy falls off for K_w/K_f less than one, just as it does for the uniformly distributed load.

It is suggested that when calculating moments in the

frame, the calculated value of P be increased by 30% since the analysis underestimates the frame shear by as much as 20 to 40%.

DISTRIBUTION OF FORCES WITH TORSION

There is horizontal torsion at any floor of a building if the line of action of the total resultant seismic force above the floor does not intersect the vertical line through the center of rigidity of the vertical bracing elements at the floor. It is usual to design a building to resist the resultant seismic force acting in the direction of each of the major axes of the structure, and either one or both of these forces may produce torsion.

The torsional moment at a floor is produced by the resultant of the seismic forces acting above the floor under consideration plus the forces acting at the floor itself. This total resultant seismic force B is in the direction of either of the two major axes of the building. The torsional moment in either direction is the force B times its eccentricity. This eccentricity is the horizontal distance from the line of action of B, to the vertical line through the center of rigidity of the vertical bracing elements at the floor, that resist forces in the direction of B. Then the floor or roof diaphragm immediately above the floor under consideration and the floor itself, will each rotate in its own plane under the action of the eccentric force B, and the bracing elements at the floor under consideration will be subjected to lateral forces due to the torsion. If the force B is eccentric in the direction of each axis of the building, the effect of each of the two torsional moments, acting separately but not concurrently, should be considered.

For certain relatively stiff types of floor and roof construction, as already mentioned, it is ordinarily assumed that the floor and roof diaphragms are rigid under the action of the torsional moment. Thus the lateral deflection of a vertical bracing element at the floor, is directly proportional to its distance from the center of rotation, which is the center of rigidity. The total applied torsional moment at any floor is then expressed as follows:

$$T = \sum (R_n e_n L_n) \qquad (18)$$

in which

T = The total applied torsional moment at the floor.
R_n = The stiffness factor of the nth bracing element, which is the lateral load that will produce unit deflection.
e_n = The lateral deflection of the nth vertical bracing element, due to the torsional moment.
L_n = The distance from the nth bracing element to the center of torsional rotation, see example.
F_n = $R_n \times e_n$. The lateral force in the nth bracing element due to the torsional moment T.
B = Total lateral seismic force that produces the torsion T.

A convenient way to evaluate the forces F_n due to the applied torsional moment T, is as follows:

$$F_n = F_{n1} \times \frac{T}{T_1}$$

in which

T_1 = The total torsional resisting moment, assuming a unit lateral deflection in the vertical bracing element farthest from the center of rotation.
F_{n1} = The lateral force in the nth element, corresponding to T_1.

A force F_n may be either positive or negative. It is positive if it acts in the same direction as the force B and negative if it acts in the opposite direction. The total maximum force acting on the nth bracing element is then the algebraic sum of the positive torsional force F_n and the force that would be produced in the element if the force B acted through the center of rigidity, that is to say, without torsion. Most building codes neglect negative torsional forces.

The bracing elements at a floor, which transmit the torsional moment through the floor to the structure below, are simply those portions of the frames or shear walls that extend between the midpoints of the stories above and below the floor under consideration. The stiffness factor R_n for any such portion of a beam and column frame, see Figure 1, may be calculated as follows.

Assume an arbitrary shear load on the element for the calculation using Equations 12 or 13 and 14, just as it would be calculated for the deflection between successive stories of the frame. In this case v_1 in Figure 1 equals v_2, for a shear transmitted from the upper to the lower story. The shear load to produce unit deflection can then be determined by direct proportion.

If the bracing elements are shear walls without openings, the R_n's may be assumed to be proportional to the moments of inertia of the walls. If there are both shear walls and frames, all of the torsional moment may be assumed to be resisted by the walls.

Accidental Torsion

In addition to the torsion that is normally computed as previously described, there are other torsional effects and influences, that are sometimes referred to as "accidental torsion." The most important of these are the following:

1. Our inability to accurately calculate relative rigidities.
2. Uncertain live load distribution.
3. Uncertainties in dead load due to variations in workmanship, materials, etc.

4. Inelastic behavior of bracing elements, such as cracking of walls, which decreases rigidities and changes relative rigidities.
5. Subsequent alterations that may be made in a building, such as the addition of walls, which not only changes the dead load but may change the position of the center of rigidity.

To allow for such effects, seismic codes often require that a building always be designed to resist an additional torsional moment. For example, the SEAOC Seismic Code has the following requirement.

> Where the vertical resisting elements depend on diaphragm action for shear distribution at any level, the shear resisting elements shall be capable of resisting a torsional moment assumed to be equivalent to the story shear acting with an eccentricity of not less than five percent of the maximum building dimension at that level.

It should be noted that this torsion requirement applies only to buildings "where the vertical resisting elements depend on diaphragm action for shear distribution." It does not apply to a building in which each vertical resisting element provides lateral support for a certain definite tributary floor or roof area, with no required load distribution among bracing elements by diaphragm action. For example, transverse torsion is not a problem for a long building with many transverse bents each of which is capable of resisting its tributary transverse seismic load.

Example of Torsion

Consider a ten-story office building having a ductile moment resisting steel frame, concrete slab floors, Spray On fire proofing of steel members and lightweight glass and metal exterior walls. The dead weight per story is about 1820 k and the total weight of the building is about 18,200 k. Figure 4 is a typical floor plan of this building. There are four identical bracing bents, numbers 1 to 8, for resisting north to south forces and six identical bracing bents, numbers 9 to 14, for resisting forces in the east to west direction. It is desired to calculate the forces acting on these bents at the 6th floor due to accidental torsion.

The base shear V is calculated from Equation 10. In this equation the factors Z and I are both 1.0, since the building is assumed to be in an area of highest seismicity and is not an essential facility. For this type of structure K is 0.67. The pe-

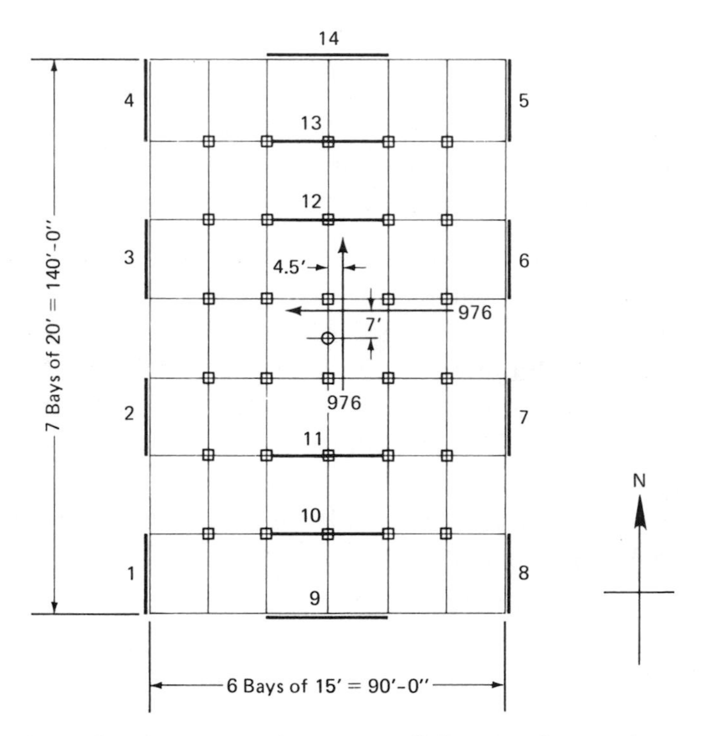

Figure 4. Typical floor plan of 10 story building showing bracing elements 1 to 14 and lateral forces producing accidental torsion.

riod of vibration may be taken as $10 \times 0.10 = 1.0$ second and for this period the value of C calculated from Equation 11 is 0.067. It will be assumed that the period T_s of the site has not been determined, in which case S must be assumed to be the maximum of 1.5. Substituting these values in Equation 10;

$$V = ZIKCSW = 0.67 \times 0.067 \times 1.5 \times 18{,}200 = 1200 \text{ k}$$

This base shear V must first be distributed to the floors by means of Equation 9. The total lateral load acting above and including the sixth floor is now found to be 976 kips. The accidental torsional forces acting at the sixth floor are then as shown on Figure 4. The center of rotation and center of rigidity is at the center of the floor plan.

The greatest accidental torsional moment is produced by the east to west force and is;

$$T = 976 \times 0.05 \times 140 = 6832 \text{ kip ft}$$

The bracing elements numbers 1 to 8 at the sixth floor are all the same and as shown in Figure 6. The deflection Δ of the element under the action of any arbitrary shear force of 12 kips, from Equation 13 is;

$$\Delta = \frac{1}{12 \times 30 \times 10^6}\left[12000\,\frac{144^2 \times 240}{4090} + \frac{117^3}{2}\left(\frac{6000}{1670} + \frac{6000}{1670}\right)\right] = 0.057$$

The stiffness factor R is then $12.0/.057 = 210$.
The direct load without torsion is then $976/8 = 122$.

The bracing elements numbers 9 to 14 at the sixth floor are also all the same and as shown in Figure 5. The deflection Δ of the element under the same shear force, from Equa-

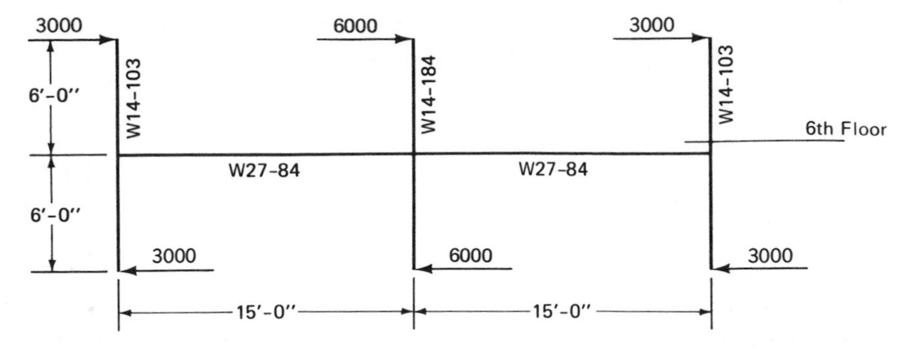

Figure 5. Typical 6th floor bracing element, numbers 9 to 14 inclusive.

tion 12 is;

$$\Delta = \frac{1}{24 \times 30 \times 10^6}\left[12000\,\frac{144^2 \times 180}{2830} + 117^3\left(\frac{6000}{2270} + \frac{6000}{2270}\right)\right] = 0.035$$

The stiffness factor R is then $12.0/.035 = 340$.
The direct load without torsion is then $976/6 = 163$.

The calculation of the accidental torsional forces acting on bracing elements numbers 1 to 8 and 9 to 14 due to the east to west seismic force is tabulated in Figure 7. The ec-

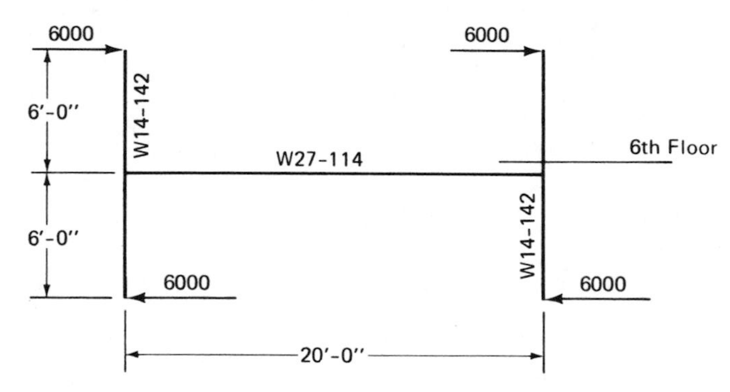

Figure 6. Typical 6th floor bracing element, numbers 1 to 8 inclusive.

Bent	L_n	R_n	e	F_{n1}	$F_{n1} \times L_n$	$F_n = F_{n1} \times .053$	$F_n \times L_n$	Total Load: No Torsion	Total Load: With Torsion
1	45′	210	0.643	135	6100	7.15	321	0	7.15
2	"	"	"	"	"	"	"	"	"
3	"	"	"	"	"	"	"	"	"
4	"	"	"	"	"	"	"	"	"
5	"	"	"	"	"	"	"	"	"
6	"	"	"	"	"	"	"	"	"
7	"	"	"	"	"	"	"	"	"
8	"	"	"	"	"	"	"	"	"
9	70	340	1.0	340	23,800	−18.0	1258	163	145
10	50	"	0.715	243	12,200	−12.9	644	"	150.1
11	30	"	0.430	146	4380	−7.7	230	"	155.3
12	30	"	0.430	146	4380	+7.7	230	"	170.7
13	50	"	0.715	243	12,200	+12.9	644	"	175.9
14	70	"	1.0	340	23,800	+18.0	1258	"	181.
					$T_1 = 129{,}560$		$T = 6832$	"	

$T = 976 \times 7 = 6832 \qquad \dfrac{T}{T_1} = \dfrac{6832}{129{,}560} = 0.053$

Figure 7. Table of torsional calculations.

centricity of 7 feet 0 inches may be on either side of the rotation center. If it is on the south side, instead of on the north side, as shown on Figure 7, the plus forces would be for elements numbers 9, 10 and 11 instead of for numbers 12, 13 and 14 as shown.

Notice that the force exerted by a bracing element is in the plane of the element. Therefore, the torsional moment exerted by the element is this force times the distance from the center of rotation to the plane of the element. This distance being measured at right angles to this plane. Elements 1 to 8 accordingly all have the same effective distance of 45 feet from the center of rotation, as shown in Figure 7. In other words elements 1 to 4 are all in the same plane.

The torsional forces due to the north to south seismic force should also be calculated. Although the eccentricity is less, only 4.5 feet, the bracing elements 1 to 8 will then be subjected to direct load plus torsion. This combination did not occur for the east to west force.

Dynamic Torsion Action

The method described above for calculating seismic torsion effects involves strictly static forces. Actually, torsional vibrations will occur and these may add significantly to the horizontal accelerations. This may produce severe effects in a building if the eccentricity is excessive, as it was in the case of the J. C. Penney Building in Anchorage, Alaska, as described in Chapter 5.

The Seismology Committee of the SEAOC did not ignore the problem of dynamic analysis. Some consideration was given to a percentage increase or magnification factor, which would be applied to torsional moments calculated on the usual static basis. Finally, it was decided to include dynamic torsion effects in the so-called accidental torsion requirement.

Up to this time, there has been no proposed design method

of analysis of torsional vibrations of a building structure. Aside from the accidental torsion requirement, the best that can be done is to reduce torsion to a minimum, by designing buildings that are symmetrical in regard to both floor plan and bracing. This is an architectural as well as an engineering problem.

DIAPHRAGMS

The function of a diaphragm in a building is to tie the structure together and distribute lateral forces to the vertical bracing elements. Diaphragms fall in two general categories, depending on the way in which they distribute lateral forces. The so-called rigid diaphragm, as its name implies, is assumed to be rigid under the action of the distributing forces in its own plane. The other type of diaphragm is classified as flexible.

The rigid diaphragm acts essentially as a flat plate that transmits lateral loads to the vertical bracing elements in proportion to their relative rigidities, either with or without torsion, as has already been described. In other words, the diaphragm is so stiff that it may be considered as rigid compared to the more flexible vertical bracing elements. The diaphragm is subjected to bending and shear, due to forces acting in its own plane.

A diaphragm that is classified as flexible, is assumed to distribute the horizontal forces by acting as a simple or continuous beam, spanning between supports at the vertical bracing elements. In this case the vertical bracing elements are considered to be rigid compared to the more flexible diaphragm. In some cases, for quite small span to width ratios of the diaphragm, it is desirable to assign loads to the vertical bracing elements by tributary areas, instead of in accordance with continuous beam reactions. This is because the deflection of a short and deep beam is mainly due to

shear, not bending. Torsion is not a factor for a flexible diaphragm.

A diaphragm should be rigid or it should be flexible. Any condition between these limits, when both diaphragm and vertical bracing elements have the same order of flexibility, presents a very complex problem of load distribution. Consequently, for design purposes, either the rigid or the flexible method of load distribution is usually employed.

Excessive diaphragm deflection may damage the building. This is particularly true when the diaphragm provides lateral support for masonry or concrete walls, which are relatively stiff. Cracks in such walls due to flexure introduce planes of weakness for forces in the opposite direction, that is to say, for forces parallel to the wall. This would reduce the effectiveness of the wall as a shear wall. If a diaphragm is classified as rigid, there need be no concern about damage to the building due to diaphragm deflection.

A committee of the Structural Engineers Association of Southern California developed the following equation for calculating the maximum permissible lateral deflection of diaphragms providing lateral support for masonry or concrete walls.[18]

$$d = \frac{h^2 f_c}{.01\ Et}$$

where

d = Maximum allowable deflection in inches.
h = Wall or pier height between horizontal supports in feet.
t = Wall thickness or effective pier depth in inches.
E = Modulus of elasticity of wall or pier material.
f_c = Allowable compressive stress in wall material in flexure, in pounds per square inch.

Concentrated forces, either loads or reactions, must be transmitted into the diaphragm so that it will not be over-

stressed in either tension or shear. For example, along the edge of a diaphragm, reaction shear forces may be distributed into the diaphragm through continuous reinforcing or through a continuous edge member, sometimes called a drag member which will resist tension and compression. The diaphragm shear is transmitted continuously into the edge member, which is connected to the vertical bracing elements along the edge. This avoids high shear stress concentrations in the diaphragm at the bracing elements. A load normal to the edge of a diaphragm, such as from a wall anchor, should be transmitted through a connecting member that extends a considerable distance into the diaphragm. The force is then taken off in shear on each side of the connecting member. This eliminates the high tensile stress that would occur in the diaphragm if the anchor was simply connected to the edge of the diaphragm.

Reinforced Concrete Slab

In every respect a poured in place reinforced concrete slab is the best diaphragm. This may be either beam and slab, in which case the beams prevent slab buckling, or it may be a flat slab or concrete plate construction. A slab can be designed for strength and stiffness, so that it will definitely fall in the rigid category. Where there are special stress conditions reinforcing bars may be added in the slab, for example, close to shear walls, at openings, or at reentrant corners. Any lateral force, either load or reaction, can easily be transmitted into the diaphragm through bond with the concrete.

Observation of earthquake damage to concrete floor slabs indicate that there are critical areas close to vertical bracing elements, particularly shear walls. A report by the California Institute of Technology (Reference 9) states that at the Indian Hills Medical Center Building (see Figure 15) after the San Fernando earthquake, "slab cracking was pronounced in regions close to some of the shear walls." The same type of slab cracking was also observed by Hanson and Degenkolb, in the 13 story Bahia del Mar Apartment building, in Caracas, Venezuela, after the 1967 earthquake (Reference 5). In these floor areas, close to shear walls or other vertical bracing elements, there are concentrated reaction forces that may produce slab cracks due to direct tension or shear. As already mentioned, a general requirement is that such reaction forces must be distributed into the diaphragm.

Steel Deck

Probably the next best diaphragm is a steel deck, which is often used in multistory steel frame buildings. There must be a concrete fill and topping on the steel deck, if it is to be considered as a rigid diaphragm, otherwise it must be considered as flexible. A steel deck is either a single corrugated steel sheet or a corrugated steel sheet welded to a flat sheet. It is shop fabricated in units generally 24 inches wide and of a length that will span continuously across the spaces between two or more supporting beams. These units are fastened together in the field by welding or so-called button punching at intervals along the seams between units. The deck is welded to the members of the steel frame to prevent buckling and to transmit shear between the units and to the frame. The deck is assumed to resist shear only and flexure or bending of the diaphragm in its own plane is resisted by steel beams that act as chord or flange members to which the deck is welded. Since the diaphragm shear is delivered to the top of a beam, there is some question as to how much of the beam should be considered as effective in providing chord section area. A common assumption is that the top flange of the beam is effective.

Comprehensive tests of eight large steel frame, steel deck diaphragms (18′ × 54′ and 16′ × 48′), both with and without concrete fill, were made and reported in 1963 by S. B. Barnes and Associates, Consulting Structural Engineers, of Los Angeles, California. The decks were tested in the horizontal position with two equal horizontal third point loads applied by jacks. On the basis of these tests, working shear stresses in pounds per foot width of deck, have been established and also the values of a so-called flexibility factor. This factor is simply the shear deflection of a 12-inch length of deck under a transverse shear load of one pound per foot width of deck. The lateral load tests of the deck did not always give well defined yield points. Accordingly, the recommended working shear stresses are based on a factor of safety of three, relative to the ultimate strength.

Steel Deck Without Concrete Fill

The strength and deflection of a steel deck diaphragm without concrete fill, depends on the following.

1. The span and width of the diaphragm.
2. The thickness and physical properties of the steel sheets. Sheets of 20 gauge give roughly half the shear strength and twice the shear deflection of 16 gauge sheets.
3. The width and length of the deck units. A 12-inch wide unit gives about the same shear strength but about twice the shear deflection of a 24-inch wide unit.
4. The type and spacing of the connections along the seam between the units. Welding gives more shear strength and less shear deflection than button punching.
5. The ratio of the span of the deck between supporting beams, to the total length of a deck unit. The smaller this ratio the less the shear deflection. Greater span is less shear strength.
6. The number and size of welds between the ends of each unit and the supporting beams.
7. The section area of the steel chords or flanges, or more accurately, the moment of inertia of the chord areas about an axis at the center of the diaphragm.

The deck consisting of a corrugated sheet and a flat sheet, should be installed with the flat sheet downward. The other type of deck, consisting of a single corrugated sheet, is not nearly as effective as a diaphragm, it develops more than three times as much shear deflection.

The working shear stress and shear flexibility factor, are determined from 2 to 6 inclusive of the foregoing list. Item 7 determines the strength and deflection of a diaphragm due to bending in its own plane. The strength and deflection are calculated from the same Equations that are employed for any homogeneous beam, using the moment of inertia of the chord areas. As previously mentioned, the steel deck is assumed to resist shear only.

Steel Deck With Concrete Fill

In order to obtain a steel deck diaphragm that is stiff enough to be considered as rigid for the purpose of load distribution, use the strongest and stiffest steel deck without concrete fill, as previously described, and add concrete fill and topping extending at least four inches above the top of the corrugations. The lateral load tests previously mentioned included two steel deck panels with concrete fill and one with vermiculite fill. The concrete or vermiculite filled in between the corrugations and extended up above the top of the corrugations provided a continuous slab 2 inches thick for the vermiculite and 2½ inches thick for the concrete. The compressive strength of the fill was about 2400 pounds per square inch for the concrete and 280 pounds per square inch for the vermiculite. The fill had no reinforcing, although the use of 6 × 6/10-10 welded wire mesh is recommended to provide some crack control. In this system the steel deck supports the vertical load, so no fill reinforcing is required for this purpose.

These diaphragms gave far less shear deflection than the

bare steel deck without concrete and a good deal more shear strength. The bending strength and deflection of a steel deck diaphragm is probably about the same with or without concrete. This is calculated in the same way in each case, with chord members resisting the flexural tension and compression.

In the three load tests of steel deck with fill, no diagonal tension cracks and no initial shrinkage cracks developed in the fill. In each of the two tests, final failure occurred by direct shearing of the fill across one end of the panel. In the third test of a deck with fill, the final failure was due to separation of the concrete from the steel deck at one end of the panel and buckling and tearing of the steel at this point. It seems apparent that the steel deck and fill form a composite section and no reinforcing of the fill is necessary to resist shear forces. Even though the steel deck was formed of smooth sheets, there was excellent bonding with the concrete. The only reported bond failure in these tests was at the point of failure at the end of a panel where the steel deck was severely distorted.

Plywood

Plywood diaphragms are used in nearly all wood frame buildings that are designed to resist lateral forces. They are also used in most earthquake resistant masonry bearing wall buildings having interior wood framing. Chapter 6 describes how this type of building may be strengthened. A very important part of this strengthening work is the design and construction of plywood diaphragms.

A plywood diaphragm is considered to be flexible with regard to lateral load distribution. If it provides lateral support for masonry or concrete walls, its deflection under code lateral forces should not exceed the permissible deflection of the wall. The deflection of a blocked plywood diaphragm uniformly nailed throughout, may be calculated by use of the following equation, which appears in the Uniform Building Code.

$$d = \frac{5\,vL^3}{8\,EAb} + \frac{vL}{4\,Gt} + 0.094\,Le$$

where

d = Deflection in inches.
v = Maximum diaphragm shear due to design loads in the direction under consideration, pounds per linear foot.
L = Diaphragm length in feet.
b = Diaphragm width in feet.
E = Elastic modulus of chord material in pounds per square inch.
G = Modulus of rigidity of plywood, which is 90,000 for structural plywood.
t = Thickness of plywood for shear in inches.
e = Nail deformation in inches. Increase 20 percent for other than structural grade plywood, decrease 50 percent for seasoned lumber.
A = Section area of each chord member in square inches.

The factor e is determined from the load per nail, and equals v/N_o Nails/ft. The values of e for 8 penny nails, which are used in the usual diaphragm with ½-inch plywood, are as follows.

Load per Nail	Value of e
60	0.008
80	0.012
100	0.018
120	0.023
140	0.031
160	0.041
180	0.056
200	0.074
220	0.096

In this deflection equation, the first term is deflection due to bending of the diaphragm, in which the flexural stresses of tension and compression are resisted entirely by the chord members. The plywood resists shear only. Although the shear deflection of a plywood diaphragm can probably be calculated with reasonable accuracy, the bending deflection is likely to be less accurate, depending on the type of chord members. For example, in wood frame construction, the usual double plate chord may have considerable elongation due to slippage of the nailed splices. On the other hand, if a masonry or concrete wall forms the chord, as may be the case for a bearing wall building, the change in length of the chords may be negligible and most of the diaphragm deflection will be due to shear.

In wood frame construction, chord members and drag members are generally double 2″ × 4″ or 2″ × 6″ plates with the joints staggered several feet apart and nailed to develop the necessary tensile resistance. As already mentioned, when a plywood diaphragm is used in a building having masonry or concrete walls, the walls themselves may serve as chords, if they have sufficient tensile strength. The tensile strength of the wall should be adequate if it is reinforced with gunite. The horizontal shear from the diaphragm is transmitted to the wall through blocking or a continuous member, bolted to the wall.

Other Diaphragm Systems

There are other types of floor and roof construction that may be used as diaphragms to a more or less limited extent. This includes insulating board and poured gypsum, which are used for roofs, and precast prestressed concrete which may be used for either floors or roofs. The manufacturers have established methods of construction and diaphragm shear values for such materials when they are used as diaphragms.

Poured in place gypsum is used for roofs. The gypsum concrete is poured to a thickness of 2 to 2½ inches on hard board forms spanning two to three feet between steel subpurlins. The gypsum concrete with light weight aggregate weighs 40 to 50 pounds per cubic foot and develops a minimum compressive strength of about 500 pounds per square inch. The slab is usually reinforced with 6″ × 6″/10–10 welded wire mesh. If the details are adequate, this construction forms a satisfactory rigid diaphragm. Diaphragm shear strengths have been established by tests and working values are specified in building codes, including the Uniform Code. Shear values for bolts set in gypsum concrete have also been established, so that bolts may be used for transmitting shear forces from the slab to a masonry or concrete wall.

A common framing detail for poured gypsum utilizes inverted “T” sections for subpurlins, with the stem of the “T” extending up into the slab and the mesh reinforcing draped over it. This is not satisfactory for a diaphragm, since the slab continuity is interrupted at each purlin and the mesh placement does not provide good reinforcing. The gypsum slab should be continuous over the subpurlins, with the mesh centered in the slab.

Insulating board is used for roof sheathing, mainly in one story buildings. An insulating board diaphragm consists of four foot wide sheet laid and nailed similarly to a plywood diaphragm. However, the material is relatively soft and an inch or more in thickness, so that longer and larger nails must be used to develop adequate bearing for shear resistance. An insulating board diaphragm should be considered as flexible.

Long span precast prestressed single and double “T” members are sometimes used for floor and roof construction in masonry or concrete buildings. Diaphragm action depends on the method that is used for connecting the edges of the slabs of adjacent members. Welded inserts spaced at intervals provide the best connection. The steel inserts must be capable of resisting either shear along the joint between the members or tension across the joint. The "T" members must also be securely anchored to the exterior walls, to resist either tension normal to the wall or shear along the wall. Frequently a 2- to 3-inch thick concrete topping is poured on the "T's." If this topping is properly reinforced and bonded to the "T's" it will greatly improve the diaphragm action. However, electrical conduit and piping are frequently laid in the topping and this may seriously reduce the effectiveness of the topping for strengthening the diaphragm.

4
Dynamic Analysis

As a basis for any dynamic analysis it is necessary to have an earthquake accelerogram, either actual or simulated. Figure 8 is the accelerogram of the 1940 El Centro earthquake. As already mentioned, this has been widely used in the past for earthquake engineering investigations. The maximum acceleration was 0.33g which occurred about two seconds after the start of the record.

DESIGN BY RESPONSE SPECTRA

Single Degree of Freedom

For every earthquake accelerogram such as Figure 8, elastic or linear acceleration response spectrum diagrams, as shown in Figure 9, can be calculated. One of these diagrams shows the maximum acceleration experienced during the earthquake, by a single degree of freedom vibrating body with a given vibration period and amount of damping for that diagram. The single degree of freedom system is a rigid mass on an elastic support, such as a vertical steel cantilever beam. As long as the vibration does not stress the supporting beam beyond its elastic limit, the diagram will correctly give the maximum acceleration. If the elastic limit of the beam is ex-

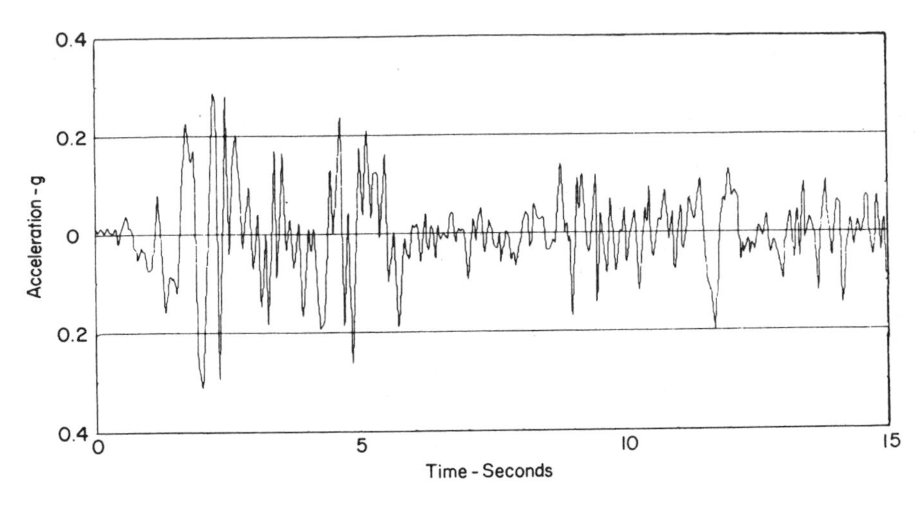

Figure 8. Ground acceleration recorded at El Centro (N-S component) in El Centro, California, earthquake of 1940.

ceeded, the response becomes nonlinear and the acceleration must be reduced by a ductility factor.

A single degree of freedom system is simulated quite well by a water tank supported on cantilever steel columns. To design this structure by the response spectrum of Figure 9, let it be assumed that the vibration period is 0.50 second and the damping factor is 0.05. Then from Figure 9, for elastic action, the acceleration at the top of the columns would be close to 1.0g and the horizontal force in pounds would be equal to the mass of the tank, W/g, times the acceleration 1.0g, which equals the weight W of the tank in pounds. Such a design would require extremely heavy columns and a massive foundation to resist overturning. It would not be practical. It must be assumed then that in this earthquake the structure will be stressed into the ductile range.

For a ductile design, assume a ductility factor of 5. Then by a commonly used method, the acceleration would be 1.0 g/5 and the structure would be designed for a horizontal

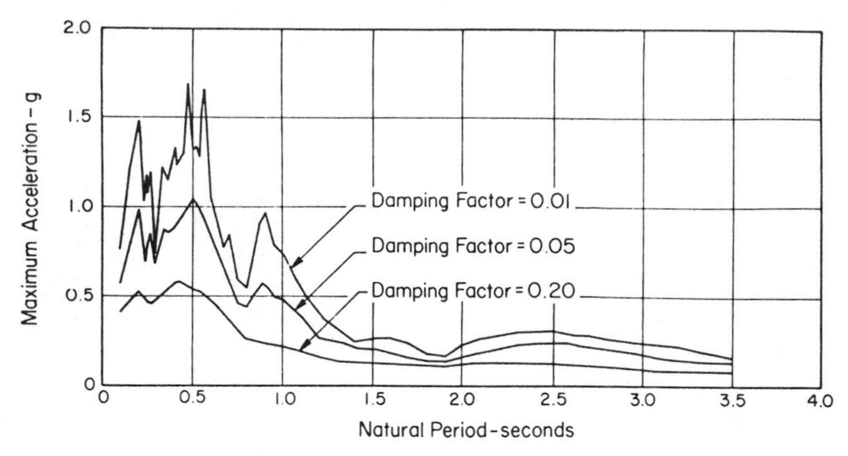

Figure 9. Acceleration response spectra for the El Centro (1940) earthquake.

force of one-fifth the weight of the tank, acting through its center of gravity.

An important uncertainty in this design is the amount of damping which must be estimated. From Figure 9 it is apparent that the amount of damping has a very large effect on the response of the structure. For example, at a 0.5 second period, if the damping factor is reduced from 0.20 to 0.05, the acceleration and therefore the earthquake force is practically doubled. Also, because of the irregularities of the response spectra diagrams as shown in Figure 9, a relatively small change in the calculated vibration period of a structure, may give a large difference in the acceleration. From Figure 9 it is seen that the greatest difference in this respect would be for short periods and small damping.

At this point it may be well to explain that the damping factor simply is the fraction (or percentage) of the so-called critical damping. If the structure has critical damping there is no vibration at all. In this condition if a single degree of freedom mass is pulled back to its maximum elastic deflection and released, it will come to rest in its undeflected position with no overrun.

Multidegree of Freedom

Response spectra may also be used for the design of a multi-degree of freedom system such as a multistory building, although this application is much more complicated. A multistory building has one degree of freedom for each story and it may vibrate with as many different mode shapes and periods as it has degrees of freedom. There is a base shear associated with each mode of vibration. For elastic action of the building structure, this base shear may be determined by multiplying an "effective mass" by an acceleration read from the response spectrum, for the period of that mode and for the assumed damping. The base shear as determined for elastic action must then be reduced for a ductile design, just as it was for a single degree of freedom structure.

The procedure described above for determining the base shear for each mode of a multidegree of freedom structure is the same as that for determining the base shear for a single degree of freedom structure except an effective mass is used instead of the total mass. This effective mass is a fraction of the total mass, which is greatest for the first or fundamental mode and becomes progressively less for the higher modes. It is a function of the actual mass at each floor and the deflection at each floor and can be expressed in a fairly simple equation. The mode shape must therefore be known in order to compute the effective mass.

The distribution of the base shear as lateral forces acting at the floors throughout the height of the building, is proportional to the lateral displacement at the floors. For modes above the first or fundamental, the displacement may be in either direction and the corresponding lateral forces must be

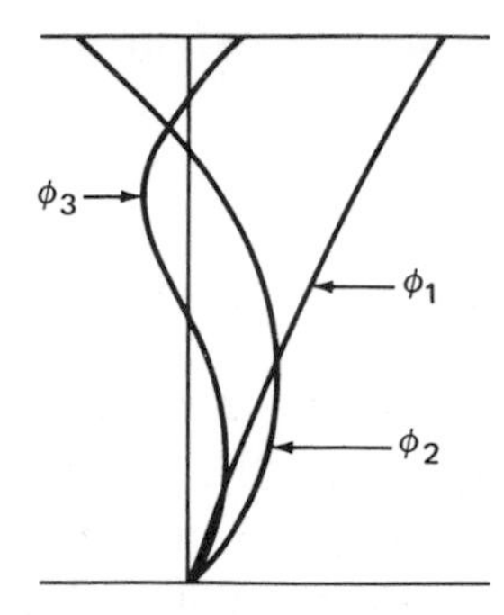

Figure 10. The first, second and third mode shapes for a 15 story building. The corresponding vibration periods are 1.25, 0.35 and 0.20 seconds.

assigned either positive or negative values (see Figure 10). This distribution requires an accurate knowledge of the mode shapes, which is also required to determine the effective mass for each mode. Mode shapes and frequencies are determined by computer, and programs are available for this purpose.

The total base shear cannot be obtained by simply adding the base shears calculated for the several modes (usually four or five) that are included in the analysis, because they are not simultaneous effects. Several approximate methods have been proposed for combining the responses of these modes of vibration to get the total effect. The exact way in which the various modes combine cannot be determined, causing a rather serious disadvantage to this method of dynamic analysis.

The response spectrum analysis as applied to multi-degree of freedom structure has now been largely replaced by a type of dynamic analysis that determines the response of the structure at short intervals of time during the entire earthquake. The chief objection to the response spectrum analysis is that the duration of the event is not taken into account. This more comprehensive method of analysis will now be described.

ANALYSIS BY MATHEMATICAL MODEL

The mathematical model must simulate the vibratory action of the building and its seismic resisting system, including torsion effects. The mass comprises the structure of the building and any essentially permanent live load. The vibration of the model, a combination of translational and torsional motion, is determined by the mass and its distribution and by the locations and stiffnesses of the bracing elements which generally consist of frames and/or shear walls. It may be difficult to simulate the stiffness of a shear wall if it has openings, particularly if the openings are irregular.

The base of the mathematical model is subjected to any desired horizontal earthquake movement such as that of the accelerogram shown in Figure 8. The base motion is applied in turn, in the direction of each of the major axes of the building. The general equation of motion for this system is integrated by computer at intervals of some fraction of a second for the entire time of the acceleration record. This gives a time history of the response, for floor accelerations, displacements and story shears and overturning moments. Damping is an input parameter and digital computer programs are available for this analysis. These are generally based on linear or elastic theory. Although nonlinear computer programs are available, these methods are time consuming and costly. If linear programs are used, the calculated response must be reduced in accordance with the ductility of the structure.

The time history of the response establishes potential problem areas within the structure, for example, points of excessive floor displacement or acceleration.

THE UNCERTAINTIES OF DYNAMIC ANALYSIS

There has been a general tendency to emphasize the approximations involved in the conventional seismic design by equivalent static forces and to represent dynamic analysis as the perfect solution. Actually it should be recognized that there are important uncertainties in a dynamic analysis.

Any dynamic analysis starts with an assumed base movement. This base movement is intended to simulate the earth movement that would actually occur at the building site during an earthquake. In our present state of knowledge it is not possible to predict the characteristics of this movement at any given site. Such a prediction must involve the so-called site effect and this is very complex as has already been described with reference to the S factor.

This complexity and the difficulty of predicting seismic earth movements is well illustrated by the results of acceleration measurements during the San Fernando earthquake. As has already been described, the Los Angeles region was unusually well covered by strong motion accelerograph instrumentation. The distribution of ground accelerations over the whole Los Angeles region was accurately measured and was found to present a complicated pattern. This situation is described as follows in a report published by the California Institute of Technology.[7]

> The complicated distribution patterns show why it is difficult to estimate closely the ground motions of a particular school site even if an accelerograph is located close by. There are significant differences between the ground motions of sites only a fraction of a mile apart. These differences are caused by variations in the propagation paths of the seismic waves, by surface and subsurface topography and by details of local geological and soil conditions. In the present state of knowledge, it is not to be expected that the motions at the various sites can be either quantitatively explained or predicted. Should another earthquake occur in a different or even the same location, it seems likely that a similarly complicated pattern of ground motion distribution would be obtained, but one which might be different from the distribution for the San Fernando earthquake.

Although the mathematical procedure for producing an Acceleration Response Spectrum from a given acceleration record is presumed to be well established, there are very important variations in the Spectrum produced by different authorities, from the same ground acceleration record. These different spectra superficially look very much alike but when they are examined in detail for use in design, the variations may result in differences of a factor of two or more in the acceleration indicated for the same vibration period and damping. In fact, maximum differences of design spectra for the same or equivalent building sites as reported by different consultants to the Los Angeles Building Department range up to a factor of four over some parts of the spectrum. This disagreement between consultants of the best reputation, regarding design spectra, introduces a serious element of uncertainty in this method of dynamic analysis.

The amount of damping in a building structure is uncertain, but this has a very important effect on its dynamic response, as is readily apparent by inspection of Figure 9. In a publication of the National Bureau of Standards entitled "Building Practices for Disaster Mitigation,"[13] there is the following statement concerning damping.

> Determination of the damping coefficients to be used (in dynamic analysis) is one of the most important and difficult steps in the seismic analysis. There are relatively few applicable test data to support an accurate estimate of the true damping of a structure. Most available test results are based on very small amplitude distortions or on component tests, and the results probably do not accurately reflect the damping that might be expected for the large amplitude motions associated with a severe earthquake. And yet, small changes in assumed damping may significantly change the calculated response of a structural system.

Another serious uncertainty is the reduction that must be made in the elastic or linear response of a building as calculated by a dynamic analysis, in order to allow for the ductility of the structure. Several methods have been proposed by different authorities for making this reduction and they give very different results. One method is to divide the calculated elastic response by the ductility factor to obtain the response of the actual structure. Since the ductility factor may range somewhere between three and six or more, it is evident that the choice of a ductility factor for any given structure will have a very large influence on the final result of a dynamic analysis. But the ductility factor of a structure cannot be calculated. Its determination is largely a matter of judgment. This was discussed in connection with the determination of K values, which are closely proportional to the ductility of the various types of structure.

The mathematical modeling of a building structure for the purpose of dynamic analysis is subject to other important uncertainties. Shear walls, or shear walls in conjunction with moment resisting frames, are now commonly used for lateral bracing of multistory buildings. But it is difficult or impossible to make satisfactory determinations of the stiffness of such bracing systems. Nonstructural partitions and filler walls can have an important effect on the dynamic response of a building and their stiffnesses are uncertainties. Also these stiffnesses will change during an earthquake due to progressive damage to these elements. Prefabricated outside wall panels are now often used in high rise buildings. Unless these panels are properly mounted so as to allow free movement of the panel relative to the building structure, they may greatly increase the stiffness of the building. This free movement must not be subject to any impairment due to improper design, poor installation or deterioration of the mount details.

The uncertainties that are involved in calculating the deflection, and consequently also the dynamic action, of a reinforced concrete frame have been described in Chapter 3.

5

Observation and Analysis of Earthquake Damage to Buildings

The observation of earthquake damage has been a very important factor in the development of methods for the design and construction of earthquake resistant buildings. In fact the basic seismic requirements of our building codes are largely based on observations by competent professionals. For example, the seismic coefficient C that appears in design equations for calculating lateral load has been primarily determined in this manner. Earthquakes provide full scale tests of structures, and they are very effective in ferreting out structural weaknesses.

In order to get useful engineering information from the observation of building damage, the earthquake should be of high intensity so that it will test well built structures. There should be modern buildings in the area with reasonably good engineering and construction. Nothing can be learned from a pile of rubble. Since no single engineer can personally observe more than a small fraction of the earthquake damage, it is essential that adequate reports by competent professionals, be available. To be useful in the United States, it is desirable that such reports of damage be made by engineers from this country, since they are familiar with our building codes and construction practices.

The following earthquakes meet the foregoing requirements; Alaska and Anchorage 1964,[8] Venezuela and Caracas 1967,[5] Manila 1968 and 1970,[10] San Fernando 1971,[9] and Managua, Nicaragua 1972.[11]

On the other hand there are many earthquakes throughout the world from which little useful engineering information can be obtained, such as: Chilean 1960 (X), Skopjie Yugoslavia 1963 (VIII), Agadir Morocco 1960 (XI), Tashkent 1966, Sicily 1968, Peru 1970, 1974 (VIII), and Guatemala 1976 (VII). The maximum modified Mercalli intensity, when available, is shown after the date for each earthquake.

The Guatemala earthquake is placed in this second category, because it was not of sufficient intensity to test the large number of multistory concrete buildings that represented United States standards. The fantastic loss of life, some 22,000 people were killed, was due to the utter destruction of the adobe dwellings of the poor.

At the sites of the other earthquakes in this category, there were either no buildings at all that represented modern construction as used in the United States, or there were relatively few. Prevalent construction was bearing wall, using either one or two story adobe, or up to six story unreinforced masonry often with lime mortar and generally very poorly anchored to the interior structure. There were a few reinforced concrete buildings. Occasionally these were well designed and constructed but often there was little or no design for seismic forces.

At Agadir some of the most spectacular structural failures were found among structures having reinforced concrete frames. Two of these failures were the eight story Immeuble Consulaire built in 1952 and the four story Saada Hotel built in 1951. These were fine looking buildings. The destruction of the Saada Hotel was widely publicized because it was one of the most luxurious hotels in the city and was popular with European and American tourists. Both buildings completely collapsed leaving simply a pile of rubble. Neither of these two buildings nor any others in Agadir were designed to resist earthquake forces. The Agadir Earthquake of 1960 was one of the most devastating local quakes of all time. Over an area of a few square miles most of the city was completely destroyed and over a third of the inhabitants were killed, about 12,000 people.

The examples of earthquake damage that are here reported and analyzed, have been specially chosen because they add to our knowledge of earthquake resistant design and they lead to suggestions for improvements in design. Commonly occurring earthquake damage that is due to failure to comply with the best current practice in design, is of less interest.

CONCERNING DUCTILITY

The SEAOC Code takes full cognizance of the importance of ductility or energy absorption in preventing the collapse of a building structure due to severe ground shaking. The need for this emphasis on ductility was dramatically demonstrated by the damage sustained by buildings at the Olive View Hospital during the San Fernando earthquake.[2]

The main building at this hospital was a very large reinforced concrete structure of five stories and a basement. The seismic lateral force was resisted by the first story columns. Figure 11 is a view of the West side of the main building immediately after the earthquake, showing the 18-inch lateral deflection of these columns. Figure 12 is a closer view of one of these columns. It was originall 27 inches square, but the outside concrete has broken off leaving a 25-inch diameter core confined by a spiral of $\frac{5}{8}$-inch diameter wires at $2\frac{1}{4}$-inch on centers. The vertical reinforcing inside the core was eight number 14 bars. In spite of this extreme deflection the column did not collapse and still supported four stories of dead load after the earthquake. This performance represents fantastic ductility. The reason is that the confinement provided by spiral

Figure 11. West wall of main building, Olive View Hospital, after the San Fernando earthquake.

reinforcing greatly increased the ultimate strength and deformation of the core concrete and also provided very high shear or diagonal tension resistance. Because of the large deformation, the longitudinal reinforcing probably was stressed well beyond the yield point and provided relatively high flexural resistance.

All of the first story columns in the main building were spirally reinforced except the corner columns. The corner columns were "L" shaped with thirteen number 18 vertical bars and number 3 ties at 16 inches on centers. These small and widely spaced ties provided no confinement for the concrete and very inadequate shear or diagonal tension resistance. As a result these columns had no ductility and broke up during the earthquake as shown in Figure 13, which is one of these corner columns. Next to the corner column the two spirally reinforced wall columns, one of which is shown

Figure 12. Column west wall main building, Olive View Hospital.

beyond in Figure 13, remained intact. These columns continued to support their load even with the additional load caused by the complete failure of the corner column.

If the first story columns in the main buildings at the hospital had collapsed, most of the people in this story would probably have been killed. As it was, there were 600 people in the building during the earthquake and only three of these people died from injuries incurred by the earthquake.

The Mental Health Building at the Olive View Hospital, a large two story concrete structure, also depended on the resistance of the first story columns for its lateral support. However all the first story columns in this building, like the corner columns in the main building, had small widely spaced ties which did not provide confinement of the concrete and gave little diagonal tension resistance, so that these columns had no ductility. The brittle failure of these columns caused the complete collapse of the first story of this building. Figure 14 shows a collapsed column at the northwest corner of this building. It is seen that this column completely disintegrated as did all the other first story columns. Fortunately this building was for outpatients and there was no one in the building at the time of the earthquake.

This building and the main building were both designed under the SEAOC Code which contains reinforcing requirements for ductile concrete columns. Unfortunately at the time

Figure 13. Tied corner column of main building, Olive View Hospital.

Figure 14. Disintegrated tied column of the Mental Health Building, Olive View Hospital, after the San Fernando earthquake. This was the north east corner of the building.

these buildings were designed, these requirements were not mandatory for the moment resisting frame in every building that depended on a concrete frame for its lateral support. This accounts for the nonductile columns in both the main building and the Mental Health Building.

Referring again to Figure 12 it is seen that there was a shear failure of a second story column directly above the spirally reinforced first story column. The spiral reinforcing did not extent into this upper column which simply had conventional ties. Entirely aside from ductility, these ties did not provide nearly enough shear or diagonal tension resistance. Column ties are sometimes thought of as simply providing lateral support for the longitudinal reinforcing bars. However they also have an important function to resist diagonal tension for columns where shear may be a factor, as it was in this instance.

SHEAR WALLS

Reinforced concrete

Shear walls are being used more and more for resisting earthquake forces, either alone or in conjunction with ductile moment resisting frames. The reason is that shear walls stiffen a building, and this reduces nonstructural damage. This was well demonstrated at Managua, Nicaragua during the 1972 earthquake.[11] There are two reinforced concrete buildings side by side, the 15 story Banco Central built in 1960 which is a frame structure and the 18 story Banco De America, which is braced by concrete shear walls in a central core. Both buildings were well designed and well constructed, within several years of each other. They were built in accordance with the United States west coast standards in force at the time of their design. The performance of these two buildings was quite different although they were subjected to the same earth movement. The structural frame of the Banco Central showed little distress but the interior was a shambles due to the violent shaking. On the other hand the Banco De America suffered no discernible nonstructural damage and sustained no serious structural damage.

It is of interest to note that the interconnecting concrete beams between the shear walls of the core in the Banco De America, sustained shear damage throughout most of the height of the building. This type of damage to members connecting shear walls in the same plane will be discussed more fully.

A common type of damage to reinforced concrete shear walls is movement and spalling along horizontal construction joints. Figure 15 is the Indian Hills Medical Center showing a horizontal crack in the shear wall at the right side of this elevation. The crack was caused by the San Fernando earthquake. Figure 16 is a close view of this same crack. Probably the spalling of concrete at the edge of the wall was due to a

Figure 15. North elevation of the Indian Hills Medical Center as seen the day after the San Fernando earthquake. A horizontal crack and spalling is visible in the shear wall at the right hand end of this elevation.

Figure 16. Closer view of the shear wall crack and spalling visible in Figure 15.

slight buckling of the vertical bars from the compression of this edge column. Diagonal tension cracks are also visible just above the horizontal crack.

Another example of this same type of shear wall damage only more severe, was sustained by the Holy Cross Hospital, a seven story reinforced concrete structure, close to the Indian Hills Medical Center. Figure 17 is a view of this building after the San Fernando earthquake. A horizontal fracture is visible at the fourth floor in the shear wall at the right hand end of the building, in this photograph. Figure 18 is a close-up view of this same fracture. At the opposite end of the building there were two shear walls, each of which had similar horizontal fractures, but they were at the third floor. Although this building had a reinforced concrete frame, it appears that the end shear walls resisted most of the transverse earthquake force.

Figure 17. Holy Cross Hospital after the San Fernando earthquake. A fracture is visible in the end shear wall.

Figure 18. Close up view of the shear wall fracture visible in Figure 17.

These horizontal fractures occurred at construction joints. Although the light weight concrete floor slab penetrated the wall along these joints, this in itself was probably not a factor in the bond failure. From Figure 18 it is apparent that there must have been a large movement at the fracture, in fact buckling of vertical wall reinforcing indicated an inch or two of vertical settlement. At the opposite end of the building there were two wall columns between the two shear walls. The extend of the movement along the fractures in these two walls was indicated by complete rupture of both these columns at the third floor, due to shear parallel to the wall.

Still another example of this same type of shear wall damage is shown in Figure 19, which is the north end of the Mt. McKinley Building in Anchorage immediately after the Alaska earthquake of 1964. The horizontal crack may be seen in the third story of the shear wall at the left side of this elevation. Figure 20 is a close view of this same crack. In this case the spalling is very extensive, indicating that there must have been

Figure 19. View of the north end of the Mt. McKinley Building, Anchorage, Alaska, after the 1964 earthquake.

Figure 20. Closer view of the northend of the Mt. McKinley Building, showing shear wall and spandrel fractures.

a good deal of motion at the crack. The two shear walls in the south end of the building, which is the same as the north end, had started to crack in the same manner in the second story.

It is evidently difficult or impossible to consistently secure an effective bond at a construction joint in a shear wall, using conventional procedures. Any sort of toothing or keying along the joint would probably not be effective, since this would reduce by half the effective section area for shear. A possible solution might be to place large diameter reinforcing bar dowels along the joint between the normal vertical reinforcing bars. Tests would be needed to determine the shear value of such dowels.

The "X" cracked spandrels shown between the shear walls in Figures 19 and 20 represent an interesting type of failure. The two shear walls tended to act together as a single cantilever beam until the spandrels failed due to the vertical shear produced by flexure. In a beam this stress is termed horizontal shear. The vertical shear in the spandrels should be a maximum at the base of the wall and decrease progressively upward to practically zero at the roof. From Figure 19 it may be seen that the "X" cracking varied in just this manner. This type of shear failure is likely to occur in any stiff member that connects two shear walls in the same plane, particularly if the member is not well reinforced for shear.

Concrete Block Walls

The San Fernando earthquake provided a very good test of the resistance of two concrete block shear walls that provided the lateral bracing at the end of the one story class room building of the boys school, at the San Fernando Valley Juvenile Facility. Figure 21 is a partial view of the end of this building after the earthquake, showing severe damage to both walls, particularly the one to the right or south side. These two identical walls were each 18 feet 8 inches long and were sup-

Figure 21. West end of the Boys School Building at the San Fernando Valley Juvenile Facility, after the San Fernando earthquake.

ported on a continuous foundation. They were separated by a door opening and were connected at the roof level by a beam that formed part of the concrete roof structure. The walls were constructed of 12″ × 8″ × 16″ open end concrete blocks with a center groove. These were laid in stack bond with all cells filled with grout. The mortar was 1:3 mix and the grout a 1:4. The reinforcing was number 4 bars at 16 inches on center each way in each of the two planes.

These walls were about 7.5 miles from the epicenter of the main shock of the earthquake and were subjected to very severe shaking. Figure 22 is a closer view of the south shear wall showing cracking off of the face shells of the blocks, over a large area. The shells also cracked off over the same area on the inside of the wall. This left only the grout cores in these areas and a small part of the block between cores. The result was a reduction of 22 percent in the area of the horizontal section of a nominal 12-inch thick wall. From Figure 21 it may be seen that the north wall was damaged in the same manner but less severely. The only other damage was some short diagonal cracks in the south wall. It may be concluded that the resistance of these shear walls was determined by their flexural strength. There was a compression failure at the bottom of the north side of each wall, produced by a powerful lurch in this direction. As noted above the compression area was considerably reduced by the cracked off shells.

The face shells apparently cracked off because of poor bond between the blocks and the grout cores. An examination of the exposed grout cores showed a small but easily perceptible separation from the block due to shrinkage. This might be prevented, and the bond improved, by using some adhesive such as epoxy in conjunction with a nonshrinking grout.

Figure 23 shows a shear failure of the concrete beam connecting the two shear walls. This is another example of

Figure 22. A closer view of the south shear wall at the west end of the Boys School Building shown in Figure 21.

Figure 23. Showing shear failure of beam connecting the two shear walls shown in Figure 21 at the west end of the Boys School Building.

the same type of failure that occurred in the spandrels between the shear walls at the north end of the Mt. McKinley Building in Anchorage, as shown in Figures 19 and 20.

COLUMN FAILURES DUE TO OVERTURNING

In Chapter 2 it was stated that the overturning requirement of the SEAOC Code was revised by omission of the reduction factor *J*. This change was made as a result of the publication of reports regarding the compressive failure of columns in several buildings in Caracas, Venezuela, during the 1967 earthquake.[5]

Because of the importance of overturning and the rather startling nature of the findings with regard to the column failures, the circumstances concerning them will be described in some detail.

Figure 24 is a view of the 18 story (and a basement) Caromay Apartment Building that is one of the buildings in which

Figure 24. East side and south end wall of the Caromay Apartment Building, Caracas, Venezuela.

column failures occurred. This is a reinforced concrete structure having a slightly curved floor plan, with eight radial three column bracing bents and one central two column bent which stops at the two elevator shafts. Ten inch deep floor joists and wide flat beams on the column lines, frame in the circumferential direction. The typical floor framing is symmetrical.

The damaging earthquake motion as indicated by the column failures was in the east–west direction, that is in the direction of the central radial bent. The spectacular damage to this building was in the basement, where four exterior columns in radial bents suffered complete failure at about the mid-height of the basement story. These were classic compression failures with cone shaped concrete spalling and outward buckling of the one inch round bars. Figure 25 shows the failure of the outside column in the central two column bent.

The authors of the report concerning these column failures (Reference 5) made a rough calculation of the overturning moments exerted on this building by the earthquake. The moments were found to be far greater than were used in the design of the structure. This design was about equivalent to Uniform Building Code Zone 3. The calculation was based on the following:

4000 pounds per square inch ultimate compressive strength of the concrete in the columns.

800 pounds per square inch dead plus live load stress for column D7

200 pounds per square inch direct stress in column D7 due to bent action overturning, in accordance with 1967 UBC Code, Zone 2

$$\text{Let X} = \frac{\text{Estimated actual overturning stress column D7}}{\text{Overturning stress column D7 by UBC Code Zone 3}}$$

$$\text{Then X} = \frac{4000 - 800}{200 \times 2} = 8.0$$

Figure 25. Showing compression failure of a column in the basement of the Caromay Apartment Building, due to the 1967 earthquake.

Evidently the UBC Code Zone 3 greatly underestimates the overturning moment. This underestimate is so great, that the relatively small influence of the *J* factor seems rather unimportant. Probably the omission of the *J* factor should be considered simply as an interim revision of the SEAOC Code, pending the development of more information relative to earthquake overturning effects. It could even be that the *J* factor should be greater than unity.

THE EFFECT OF INFILLED MASONRY WALLS

In Caracas practically all the multistory buildings have reinforced concrete frames that were designed to support all vertical loads and resist all the horizontal forces. Walls and partitions are predominantly of clay tile which are infilled on the concrete frames. Only the weight of this masonry was considered in the design of the frames, no consideration was given to the stiffening effect of these walls and partitions. As a result the distribution of the lateral load assumed in the design, which was based on the relative rigidities of the bare frames of the bracing bents, was often quite different from the actual distribution.

Bents which had these nonstructural hollow clay tile infills, received more lateral load than that for which they had been designed. The additional overturning moments probably caused many of the column failures that were observed in these buildings. There appeared to be a definite correlation between the locations of the tile walls and the column failures, particularly at the corner columns where there was solid exterior end wall.

Inspection of multistory buildings in Guatemala City after the 1976 earthquake, showed this same prevalent use of nonstructural masonry walls infilled on the reinforced concrete frames of the buildings. Here the masonry was unreinforced brick, rather than unreinforced hollow clay tile as was used in

Caracas. These infilled walls presented the same structural disadvantages that are described above, involving changes in load distribution and overstressed columns. In fact, the typical reinforced concrete frames in buildings in Guatemala City, generally did not have an opportunity to act due to resistance of these infilled masonry walls.

It should be emphasized that these unreinforced masonry infilled walls as used in both Caracas and Guatemala City, were strictly nonstructural. They could not be counted on to provide any permanent shear wall resistance. There would be significant resistance only as long as the walls remained intact. These walls were not keyed in or anchored to the concrete structure in any way.

TILT-UP CONSTRUCTION

During the San Fernando earthquake there was a great deal of damage to one story buildings of tilt-up wall construction. In one industrial area, 11 out of the total of 18 tilt-up buildings received serious structural damage. In this country probably over half of all one story industrial buildings built during the last ten years are of this type. It is therefore important to analyze the damage to these buildings, particularly since this damage was largely due to the failure of details that had been pretty well standardized.

Tilt-up construction derives its name from the fact that concrete wall units are poured in a horizontal position and then tilted up and connected by pouring concrete closures in the gaps between the units. The horizontal wall reinforcing splices in these closures. The roof construction consists of wood beams supported on the exterior walls and interior steel pipe columns, wood purlins framing into the beams on joist hangers and 2″ × 4″ rafters, generally at 24 inch on center, carrying the plywood roof deck.

The plywood roof deck is nailed to the top of a wood ledger

which is bolted to the wall. The connection of the roof to the walls is entirely dependent on this nailed joint. It is not strong enough for earthquake loads normal to the wall where edge distance of the nails is probably an important factor. This weakness is illustrated by the roof collapse shown in Figure 26, which is typical of the damaged buildings. At the top of the wall in this view, may be seen the bolted wood ledger from which the plywood roof deck has torn away. The rod braces shown, were simply installed temporarily to give the wall lateral support until the roof was reconstructed.

The typical purlin connection at the wall was simply a joist hanger on the ledger, which is good for vertical load but has no tension value. Each purlin should have a positive anchor to the concrete.

Inadequate beam anchorage to the wall was also responsible for roof failures, such as that shown in Figure 26, where the beam has pulled away from the wall. The beams should always

Figure 26. Showing collapse of a portion of the roof of a tilt-up building in San Fernando, due to the 1971 earthquake.

be framed to the concrete wall closures, where a good bolted anchor can be encased in the poured in place concrete. From a close examination of Figure 26, it will be seen that the two roof beams shown did not frame into the closures, two of which are indicated by pairs of vertical lines on the inside face of the wall. These beams were supported on the precast wall panels, where it is not possible to secure a good anchor or a good vertical support. It should be remembered that there can be no projecting steel members on the inner face of the wall panel, since this face is down on the concrete floor slab when the panel is poured.

The connection of the rafters to the wall provided only for vertical support at the ledger, there were no anchors. Every other rafter, or at least every third rafter, should have a positive anchorage to the concrete.

This earthquake also demonstrated a defect in the tilt-up type of wall construction. There was frequently a separation or lateral movement at the joints between the poured in place concrete closure and the precast wall units, whether or not there was wall collapse. These closures should be, and usually are, made with a keyed joint to the wall panel, for water tightness and also to prevent relative lateral movement normal to the wall. This should be effective. Separation is undoubtedly due to inadequate bond in the splices of the horizontal wall bars. This could be improved either by widening the closure from the usual 12 inches to perhaps 18 inches so as to give more bar lap at the splice, or by hooking the ends of the bars.

This tilt-up construction is very rapid and economical, which is the reason why it is so widely used. In order to still further speed up construction and reduce the cost, a so-called "panelized" system is often used to construct the plywood roof diaphragm. In this system the deck is built-up from rectangular panels fabricated in the shop. This construction did not appear to perform as well in the earthquake, as did diaphragms built-up entirely in the field. In the latter type of construction, the end joints of the plywood sheets are staggered, which gives added tensile strength to the diaphragm in one direction. This cannot be done with panelized construction.

TORSION

Torsion occurs under the action of earthquake forces when the center of the mass of a building does not coincide with its center of rigidity. If there is torsion, the building will rotate about its center of rigidity. This may cause large increases in the lateral forces acting on bracing elements and on other parts of the structure, in proportion to their distances from the center of rotation.

During the Alaskan earthquake the J. C. Penney Building in Anchorage was so severely damaged by torsion that it was necessary to demolish the structure. This may be considered as a classic example of what torsion can do to a building.

This building had five stories, with 10-inch thick reinforced concrete flat plate floors supported on concrete columns. Figure 27 is a plan of the building. It was braced by 8-inch thick reinforced concrete shear walls, which are shown on the plan by wide black lines.

For resistance to north to south forces, there was a shear wall the full length of the west wall and two shear walls each two bays long on the east wall, however, the southerly of these two bay shear walls extended only to the fourth floor. For north to south forces, this shear wall configuration resulted in extreme torsion above the fourth floor and severe torsion below the fourth floor. For resistance to east to west forces, there was essentially only one shear wall, which extended along the south wall of the building. There was no shear wall along the north side of the building and this resulted in extreme torsion under the action of east to west forces.

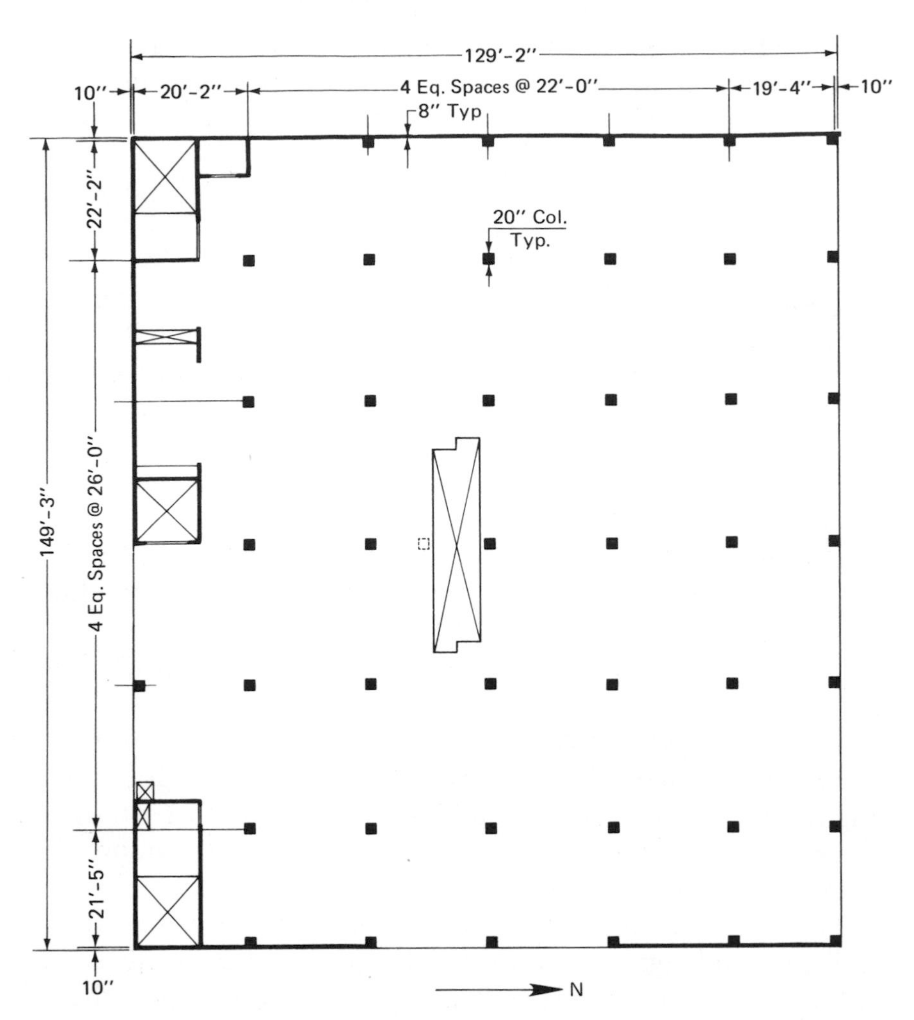

Figure 27. Floor plan of the J. C. Penney Building in Anchorage, Alaska, showing the highly eccentric shear wall configuration.

Under the action of north to south forces, the two shear walls in the east wall received very large load increased due to torsion, and both walls failed in shear, as shown in Figure 28, which is a view of the east wall and the northeast corner of the building. In this view it can be seen that the far or southerly

Figure 28. East wall and north east corner of the J. C. Penney Building, after the 1964 earthquake. This shows the failure at two horizontal joints in the shear wall near the south east corner and complete collapse of the shear wall and portions of the roof and floors at the north east corner of the building.

shear wall failed on two horizontal joints, which is a common type of failure for shear walls, as already noted. The complete collapse of the two bay shear wall and portions of the floors and roof at the northeast corner of the building, requires some explanation. This was undoubtedly the result of east to west earthquake forces increased by torsional rotation. The southerly two bay shear wall did not collapse in this manner because it was close to the center of rotation near the south wall.

At the north end of the west shear wall, a portion of the wall in the second story was thrown outward off its support on the wall below so that it dropped several feet. This was due to the same torsional action that produced the collapse at the northeast corner of the building.

Before the earthquake, the entire north and east walls of the building were covered with 4-inch thick precast nonstructural concrete wall panels extending from the second floor to the roof. Most of these panels were thrown off the building during the earthquake. Figure 28 shows a single panel that still remained in place at the south end of the east wall. It is significant that this panel was close to the south wall, where the torsional movement and acceleration due to east to west forces was a minimum.

The destruction of the precast walls and the east shear wall and much of the damage to other parts of the building was probably largely due to torsional vibrations and to increased accelerations from torsion. The conventional analysis for torsion simply gives the forces acting on bracing elements due to the moment produced by an eccentric static force. It takes no account of the torsional vibrations and the associated accelerations.

WOOD FRAME CONSTRUCTION

Wood frame construction is relatively light in weight which in itself reduces seismic forces. The average wood frame house, with plaster walls and partitions having relatively small openings, has good earthquake resistance. There were however a considerable number of badly damaged wood frame houses that were subject to very intense shaking during the San Fernando earthquake. All these houses were partial two story-like the one shown in Figure 29, which is a view after the earthquake. The part of the house on the left in this photograph was originally two story with bedrooms above a garage. During the earthquake, this portion pulled away from the one story portion (at the right) and its first story collapsed, so that the second story fell to the ground in the position shown in the photograph. During the earthquake the people in this house were in the second story bedrooms and they simply rode the second story down to the ground. No one was injured.

Figure 29. Split level house after the San Fernando earthquake, showing original second story that has pulled away from the one story portion of the building and fallen to the ground.

This failure was due to two serious defficiencies. The two story portion of the building had no lateral bracing in the front of the first story and was not tied in any way to the more stable one story portion. The garage had a 16 feet wide front door with narrow pieces of wall each side that left little space for any kind of bracing. The only attempt at bracing was a strip of plywood nailed to the studs on each side of the door which had practically no bracing value.

Figure 29.1 is a view of another split-level house that survived the earthquake without collapse of the two-story portion, but with severe cracking and deflection of the narrow pieces of wall each side of the garage door. It will be noted that the one-story roof has dropped down where it meets the two-story stud wall, but without any separation at this point.

This type of structural deficiency, large door openings

Figure 29.1. Another split-level house after the San Fernando Earthquake. The two-story portion did not collapse but there was serious damage to the first story.

with insufficient wall for lateral bracing, is not uncommon in wood frame construction. For example, Figure 29.2 shows a two-story wood frame wall having a series of wide garage door openings, with no wall at all in the first story. The result was severe lateral deflection of this story as a result of the San Fernando earthquake. However, it will be noted that the second story wall, having smaller openings, did not deflect appreciably and was undamaged except for two small cracks.

Solid wall was omitted in the first story of this building in order to give maximum space and clearance for the garages. This could have been accomplished, and the damage and severe lateral deflection could have been avoided, by using steel beams over the garage doors supported on steel

Figure 29.2. The rear wall of a large two-story wood frame apartment building in San Fernando after the earthquake showing severe lateral deflection of the first story due to lack of bracing.

columns with moment-resisting beam to the column connections. Wood framing would probably not have been adequate. As it was, there was no lateral bracing of this large apartment building in the first story of the rear wall. The lateral force here must have been resisted by interior partitions but with severe lateral deflection and probably with severe cracking.

A VERY HAZARDOUS TYPE OF BUILDING

Figure 30 shows a three story brick bearing wall building on the main street of San Fernando, immediately after the 1971 earthquake. Nothing of engineering value can be learned from this damaged building. It incorporates no earthquake resistant features. It is included here simply because it shows so drama-

Figure 30. Unreinforced brick bearing wall building on the main street of San Fernando, immediately after the San Fernando earthquake.

tically what usually happens to a traditional unreinforced brick bearing wall building, during a severe earthquake. This is the type of building construction that presents so serious an earthquake hazard in California cities and elsewhere in seismic regions.

This particular building was built in 1914. It embodied the usual structural deficiencies of this type of construction. The unreinforced brick walls with lime mortar had very poor anchorage to the interior framing. Along the street front these anchors were at 6 feet 0 inches on center at the third floor and 10 feet 0 inches on center at the roof. They were the traditional "T" anchor, consisting of a steel rod hooked into a wood floor or roof member and bent in a closed loop around an 8 inch or 10 inch long horizontal or vertical bar set in the brick masonry. Although the interior right angle hook is quite effective for connecting the anchor rod to a wood framing member, the cross bar of the "T" which is set in the wall, has very little anchorage value in lime mortar masonry. Also the anchors, such as they were, were set too far apart to give the wall proper lateral support.

In this failure the masonry of the second and third stories of the side wall shown in Figure 30, fell through the roof of the one story store building next door. The entire floor of this building was piled up with masonry and other debris when the Author inspected the building immediately after the earthquake. Since this building was on a street front, which would usually be the case for such a building in a city, the collapsed front wall fell on the sidewalk. It is this falling masonry that is so hazardous when buildings of this type are subjected to an earthquake.

6 Earthquake Strengthening of Buildings

The first requirement is to determine whether a building should be strengthened. It is safe to say that any building must be strengthened if it was designed and built before the applicable building code contained earthquake resistant design requirements. This situation often exists, since the inclusion of earthquake requirements in building codes is relatively recent, for example, 1933 in the City of Los Angeles. Even if a building was designed under a seismic code, it may have deficiencies, and this can only be determined by a review of the structural design and an inspection of the building.

If strengthening is necessary, the strengthened building structure should meet the same requirements as prescribed for a new building, relative to vertical bracing elements, diaphragms, and foundations, as described in Chapter 3.

A very important requirement for any strengthening work is that it does not in any way change the action of the structure so as to reduce its strength in some manner. For example, an added shear wall might cause torsion that did not exist before. Introduction of new vertical bracing elements or changes in the relative rigidities of existing elements, may increase the shear and bending stress in the horizontal load distributing diaphragms.

Buildings that are to be earthquake strengthened may or may not have been subjected to earthquake forces. The full scale test of a building by an earthquake may be very helpful to the engineer in planning a strengthening project. As already pointed out, an earthquake is very effective in ferreting out structural weaknesses. Whether or not a building to be strengthened has been subjected to an earthquake, the design of the strengthening should include, so far as possible, all work necessary to bring the structure of the strengthened building up to the requirements of the applicable building code. These requirements include vertical load as well as lateral load. Aside from earthquake resistance, there may be deficiencies in the design and/or construction of the original building that should be corrected so far as possible, as part of the strengthening project.

The design of the strengthening should be reviewed by the building department that has jurisdiction. This review should include not only the actual calculations, but the basic concepts, judgment factors, design assumptions, and the interpretations of the requirements of the applicable building code. This procedure utilizes the abilities not only of the designing engineer, but also those of the building department personnel, who have presumably had experience on other strengthening projects. Earthquake strengthening of buildings is both a science and an art, and it is influenced by the development of new ideas involving both design and construction.

This is the procedure that has been followed in the strengthening of California school buildings, the State Office of Architecture and Construction having jurisdiction and the applicable building code, Title 21 of the California Administrative Code. This procedure has been eminently successful, as witnessed by the very slight structural damage sustained by earthquake strengthened school buildings during the San Fernando earthquake.[7] At the time these school buildings were strengthened they had never been subjected to an earthquake, and the San Fernando earthquake tested the strengthening work that had been done.

In California, most of the strengthening of buildings to resist earthquakes has been done on school buildings. As mentioned in Chapter 2, the State Legislature passed the Field Act in 1933, which gave the State Office of Architecture and Construction jurisdiction over the structural design and the construction of school buildings. Rules and regulations originally called Appendix A and now called Title 21, were adopted to govern this design and construction. None of the school buildings that were built prior to 1933 met the new earthquake resistant design requirements. Accordingly, a law was ultimately enacted that required such school buildings to be either strengthened so as to meet the new requirements, or be abandoned. A time limit was established for the accomplishment of this strengthening, but it became necessary to gradually extend this time because of the shortage of class rooms and the high cost of the earthquake strengthening.

The fact that almost all the earthquake strengthening of buildings in California has been confined to schools, simply means that reduction of this hazard has been given first priority. It is common knowledge that buildings having inadequate resistance to earthquakes are very numerous and present a serious hazard not only in California but in all geographical regions that are subject to earthquakes. Recently the General Manager of the Los Angeles City Department of Building and Safety stated that there are approximately 14,000 buildings in Los Angeles that are considered hazardous. These buildings were built prior to 1933, when the first earthquake provisions were incorporated in the City Building Code. The cost of replacement or strengthening of these buildings has been estimated to be between four and five billion dollars. This gives some idea of the magnitude of this problem.

An ordinance dated Oct. 1976, has been proposed, which

would amend the Los Angeles Municipal Code, to require posting of earthquake hazardous buildings. For the purpose of this ordinance any building that meets the following description shall be deemed as earthquake hazardous.

1. The building was constructed prior to October 6, 1933, which was the date when seismic requirements first appeared in the Code.

2. The building has unreinforced masonry walls which provide vertical support for a floor or roof and when the total load is over 100 pounds per linear foot. Masonry walls shall be considered as unreinforced if the reinforcing steel is less than 50 percent of the minimum steel required by the present building code.

The owner of the building, determined by the Los Angeles City Department of Building and Safety to be earthquake hazardous, shall post a sign in a conspicuous place on the building stating that the building is unsafe for occupancy during a moderate or severe earthquake. This sign may be removed only if the building is brought into conformance with the current horizontal force requirements of the building code.

The foregoing is concerned with the prevention of so-called structural damage. That is to say, damage that will reduce the ability of the building to resist vertical loads and lateral forces, or which will actually endanger its stability. It is also important to prevent nonstructural damage. This includes broken glass, damage to plumbing, elevator damage, plaster cracking, falling of ceilings and light fixtures, moving or overturning of furnishings and equipment, and breaking loose and falling of exterior wall panels, parapets, and ornamentation. This type of damage is hazardous to persons both inside and outside the building. It is less expensive to make such alterations to a building as are necessary to reduce or prevent nonstructural earthquake damage, than it is to repair such damage after it has occurred.

LEGAL REQUIREMENTS FOR STRENGTHENING THE BUILDINGS WITHIN A CITY

The foregoing describes what may be termed the ideal earthquake strengthening procedure. The result is a strengthened building that has the same earthquake resistance as a new building. If the design is in accordance with a good modern building code, such as the SEAOC Seismic Code, this reduces to a minimum both property loss due to an earthquake and hazard to human life. In recent years, several cities in California have enacted ordinances that require strengthening or demolition of buildings that present an excessive earthquake hazard to persons and property. In each case an effort was made to secure an ordinance that does not create too great a hardship on property owners and still does not involve too great a risk to the general public. Thus these ordinances are a compromise.

In order to do this it is necessary to have somewhat more flexible earthquake strengthening requirements. This was accomplished by introducing two factors in addition to earthquake resistance for determining earthquake strengthening requirements, namely, "average human exposure" and the life of the building to demolition or strengthening.

Average human exposure is expressed as an equivalent number of people exposed 24 hours a day every day of the year. The calculation of human exposure includes not only the persons working or residing in the building, but also persons who are outside of the building but so located with respect to it as to be endangered by structural failure due to an earthquake. Persons walking across on the sidewalk in front of the building are in this second category. Each person is exposed for their transit time.

For example, consider a building in which 100 persons are employed steadily for 40 hours during the week. The total man hours for persons in the building is then 4000 per week. Suppose also 200 people walk in front of the building each day and that the transit time for each person is 30 seconds. This represents 12 man hours per week. The total man hours per week is then 4012, and if this is divided by 168 hours in the week, the average human exposure is 25 persons.

The first step in application of a strengthening ordinance is inspection and grading of all buildings in the city by City Building Officials in accordance with their earthquake resistance and in proportion to the average human exposure that they present. A priority is followed for this inspection, taking the most hazardous type of construction first. This would be buildings that utilize unreinforced masonry bearing walls.

The Long Beach ordinance, enacted in 1971, establishes four classifications of human exposure, namely, 1 to 9 persons, 10 to 99, 100 to 999, and 1000 or over. The required earthquake resistance increases by a factor of 2 to 3 as the human exposure increases between these limits. A building that is graded by the City Officials as carrying excessive hazard may perhaps be rendered acceptable without strengthening if the average human exposure is reduced. This can be done either by changing the occupancy or by owner imposed reductions in the actual occupancy such as would be achieved by closing off one or more of the upper floors of a building. This may give the owner an attractive option. Of course the change would increase the risk for people in the building.

The Long Beach ordinance has a table showing the earthquake resistance requirements for buildings having various life spans to demolition or strengthening, varying from 2 years to 80 years. This gives the owner of a building that is graded as excessively hazardous, the option of allowing the building to stand for a definite maximum period with little or no strengthening, or doing an immediate more extensive strengthening job. For example, if the life of a structure is reduced from say 40 years to 10 years, the required earthquake resistance is reduced by about half.

The ordinance for strengthening adopted by the city of Santa Rosa in 1971, establishes strengthening requirements for three life term categories, as follows: for buildings to be used continuously over an unlimited term, for buildings to be used for a term not to exceed five years, and for buildings that will be demolished or fully strengthened within one year. This is more conservative than the Long Beach ordinance.

It is difficult to assess these strengthening ordinances without actually experiencing the difficulties attendant on their enforcement. The good judgment of the engineer or engineers who prepared these ordinances is very important, since this involves so many judgment factors, such as the establishment of modified seismic coefficients, the effect of the life of the building etc. The SEAOC Seismic Code on which these ordinances are based, makes no mention of human exposure. It simply requires an increased earthquake resistance for certain buildings that would be essential after an earthquake, for example, hospitals and fire houses. Neither does it make any mention of structure life. It would seem that the preparation of strengthening ordinances should not be the work of only a few engineers; it would be much better to have an organization such as the SEAOC or the EERI prepare a more or less standard ordinance for strengthening.

GUNITE

Gunite, also called Shotcrete, is pneumatically applied concrete. Two different methods of operation are employed.

In the first method the wet concrete mix, usually Ready Mix, is pumped through a hose to a nozzle at the point of application. Compressed air is injected near the nozzle forming a spray that is directed against the surface. Concrete applied by this method is termed *Shotcrete* by the equipment manufacturers and is called *wet mix*.

In the second method, the concrete may also be Ready Mix, except that it is delivered dry on the job. It is then placed in a gunite machine from which it is delivered through a hose by compressed air to a nozzle at the point of application. Water is injected near the nozzle forming a spray. Concrete placed by this method is termed *gunite* by the equipment manufacturers and is called *dry mix*.

For lengths of hose up to 100 feet, the required air pressure at the gunite machine for delivering the dry gunite mix to the nozzle, should be 45 psi or more. Where the length of hose exceeds 100 feet, the pressure should be increased 5 psi for each additional 50 feet of hose. Water injected at the nozzle should have a pressure at least 15 psi greater than the air pressure at the nozzle.

It seems preferable to use the term *gunite* because this is understood by the equipment manufacturers to mean the so-called *dry mix*, with which the author is personally familiar. This method is better than the wet mix for use in strengthening or repairing buildings, where the concrete is placed around reinforcing steel, because the slump or stiffness of the gunite can be controlled by the nozzle man by adjusting the amount of water so as to get the best placement. The nozzle man should have several years experience.

The usual gunite mix is portland cement, sand and ⅜-inch maximum size aggregate. Rebound during placement is excessive with larger aggregate and too much material is thereby lost. The usual 28-day compressive strength is 3000 to 4000 psi. Two 6″ × 12″ test cylinders of gunite are generally required to be made by each nozzle man on the project during each day of operation. These test cylinders are made with small mesh wire fabric as a form with the same nozzle tip, air pressure and hydration as is used for the gunite in the structure at the point where the cylinder is made.

Gunite is widely used for strengthening or repairing concrete and masonry structures. Experience has shown that gunite bonds well with brick walls in the repair of masonry buildings. In fact, gunite bonds to almost any surface—hard, soft, wet or dry, and even to polished steel. It is likely that gunite, that is, the dry mix, bonds even better than the wet mix, because the velocity of the concrete from the nozzle is higher.

Placing Gunite

Masonry or concrete surfaces should be thoroughly cleaned and roughened by sand blasting before the gunite is used. Before the application of gunite the concrete and masonry should be thoroughly cleaned of all debris, dirt and dust, and wetted, but not so wet as to overcome suction. Any deposits of loose sand or rebound should be removed before guniting. Generally the guniting operation is performed by two people; one handles the gunite nozzle while the other uses a compressed air jet, which precedes the nozzle to blow out rebound and sand that may have lodged behind reinforcing bars. Sagging of gunite after application should be avoided by not attempting to build up too thick in one operation. Should rebound pockets, sags, sloughing or other defects occur in the work they should be cut out and replaced.

At a construction joint the film of laitance which forms on the surface of the gunite should be removed within two hours after application by brushing with a stiff brush. If the film is not removed within two hours, it should be removed by wire brushing or sand blasting. Construction joints should be thoroughly cleaned with air and water.

STRENGTHENING OF MASONRY WALLS

Probably the greatest hazard in regions subject to earthquakes, is due to unreinforced masonry bearing wall buildings with interior wood framing. The walls of such buildings are likely to fall out, as happened to the building shown in Figure 30. In the United States the walls of these buildings are almost always brick, and they are constructed with either lime mortar or portland cement mortar. The old buildings are generally built with lime mortar as was the case with the building of Figure 30, which was built in 1914. All such walls having lime mortar must be completely strengthened to resist earthquake forces. If the walls are constructed of a good quality of portland cement mortar, they should still be strengthened, but the strengthening requirements may be somewhat less stringent. Decisions in such cases must depend on tests of the masonry and mortar and on engineering judgment or on building code requirements.

Exterior Guniting

The gunite reinforcing on the exterior of a masonry wall consists of a continuous 3- or 4-inch thick slab on the face of the wall and vertical ribs, integral with the slab, cut into the wall at intervals. The width of these ribs or chases cut into the wall, should be at least 50 percent greater than their depth, for good guniting. Sometimes if the masonry wall is brick of a thickness of 13 inches or more, the outer course of brick is removed. The gunite is then applied to a thickness of about 4½ inches, so as to bring its outer surface to the original face of the brick. The slab is generally reinforced with number 3 or number 4 vertical and horizontal bars. The ribs are reinforced with vertical bars and horizontal ties, like a column. The slab is designed to span horizontally and transmit the seismic load normal to the wall to the vertical ribs. These ribs in turn span vertically and transmit the load from the slab through anchors, to the interior structure of the building at the floors and roof. The spacing of the ribs is usually about 6 or 8 feet, and there should be a rib close to the jamb of each window and door opening in the wall. This spacing of ribs is determined either by the strength and stiffness of the slab for transmitting the loads normal to the wall to the ribs, or it is limited by the buckling strength of the slab for resistance to shear forces parallel to the wall. Stiffness of the slab and buckling strength is usually controlled by limiting the maximum span to the thickness ratio of the slab. All strength calculations are based on the gunite only, without any regard to any possible composite action of gunite and masonry.

For a thick wall, plugs of gunite are often extended inward from the slab spaced at intervals between the ribs, so that the inner units of the masonry will be securely held in position by contact with gunite. The masonry wall should be self-supporting during guniting operations. Accordingly horizontal cuts or chases in the wall should be avoided.

If a brick wall is constructed of a good quality of Portland cement mortar, the gunite slab may sometimes be omitted and the strengthening simply consists of the gunite ribs. The masonry must then be strong enough to span horizontally between the ribs. Such masonry has considerable tensile strength in flexure in the direction of running bond. This strength will be about twice the tensile strength across a bed joint. If the gunite slab is omitted, the masonry must also have sufficient shear and diagonal tension resistance to act as a shear wall. It is unlikely that the gunite slab can be omitted if the wall has window and door openings.

As already mentioned, for forces normal to the wall, the wall is connected to the inner structure of the building by wall anchors that extend from the floors and roof into the gunite ribs (see Figures 31, 32 and 33). The method used for transmitting forces parallel to the wall from the floor or roof to the wall, depends on the type of diaphragm that is used (see Figures 31, 32 and 33). At the bottom of the wall, the vertical bars in the gunite ribs are doweled into the concrete footing of the masonry wall and the vertical bars in the gunite slab are also doweled into the footing.

An Example of Guniting

Figure 30.1 shows gunite work in progress on the wall of a classroom in a one-story school. Like all class rooms, this wall has a series of large windows which have been covered for protection. The only gunite is in ribs between the windows. The man on the left is handling the compressed air jet for blowing out rebound and sand. The nozzle man is in the middle, and the man on the right is trowelling a smooth surface on gunite that has already been placed. In this view the largest hose, probably about 2 inches in diameter, carries the dry gunite mix under air pressure and the smaller hose alongside it is carrying water, also under pressure, to be injected at the nozzle. One of the ribs is in the process of being gunited and another rib on the left of it shows exposed reinforcing steel since it is only partially gunited. The rib on the extreme right, which is being troweled, is wider than the other two since the windows here have a wider spacing.

Figure 30.1. Gunite work in progress on a one-story school building.

This is an interesting example of the strengthening of a one-story wall having large closely spaced window openings. This can be done very well with gunite ribs which are doweled into the concrete foundation and bonded at the top to a continuous reinforced concrete bond beam or lintel over the windows.

Interior Guniting

If there is no access to the outer face of a masonry building wall, for example, if the building is in contact with or close to another building, then the wall guniting must be done from the inside of the building.

The vertical gunite ribs are cut into the walls between joists and are continuous from the first floor to the roof. The portion of the rib that passes through the floor must be poured concrete since it is not readily accessible for guniting (see Figure 34. Where the joists are parallel to the wall, the ribs may be located wherever desired, since there is no joist interference. The gunite slabs are continuous from a floor to the bottom of the joists of the floor above (see Figure 34). In this case the slab is always gunited on the face of the wall. Removal of a course of brick would reduce or remove the end bearing of the floor joists. The guniting details and the method of design are the same whether the wall is gunited from the inside or the outside.

When the wall is gunited from the inside, there is no gunite slab between the top of the floor and the bottom of the floor joists. This should have no significant effect on the resistance of the wall to forces normal to it, since this narrow strip of unsupported masonry certainly will not break out. The resistance of the wall to forces parallel to it is determined by the shear resistance of this strip of unsupported masonry, plus the shear resistance of the concrete portion of the gunite ribs. Therefore the concrete portions of the ribs should have closely spaced ties around the vertical bars. This total shear resistance should generally be adequate, but it is an uncertainty. Accordingly a wall should be gunited from the outside wherever possible.

Gunite for strengthening masonry-bearing walls has been used on many large buildings up to 4 or 5 stories in height. Gunited walls in conjunction with proper wall anchorage and adequate horizontal load distributing diaphragms at floor and roof, will give an excellent building capable of resisting a severe earthquake.

Other Uses of Gunite Strengthening

Gunite is not only used to strengthen masonry walls, it is also used on concrete walls. This was done at the Indian Hills Medical Center building to strengthen concrete shear walls that had been damaged by the San Fernando earthquake. An 8-inch thickness of reinforced gunite was added to the existing 8-inch thick concrete wall, the cracks had been repaired with epoxy (see Figures 15 and 38).

Gunite may also be used to enlarge and repair a concrete column or beam, for strengthening purposes. Forms for columns should be open on two sides where possible, to facilitate shooting. The open corner is established by a wire ground to which the two sides are rodded. Beams should have only a soffit form with a vertical form of wire mesh near the center, so that gunite can be shot from both sides.

CORRECTION OF DIAPHRAGM DEFICIENCIES

A *diaphragm* has been defined as a horizontal element that ties the structure together and distributes lateral forces to the vertical bracing elements. This means that the diaphragm is subjected to bending and shear in its own plane, since it acts as a beam loaded by horizontal earthquake forces and spanning between shear walls or other vertical bracing elements. Very important diaphragms are the wood floor and roof structures in masonry-bearing wall buildings. These structures are almost always inadequate to perform the functions of a diaphragm. Gunite strengthening of the walls of a masonry bearing-wall building serves no useful purpose unless the diaphragms of the building are adequate.

If a diaphragm functions adequately it must meet all of the following requirements. (1) The diaphragm itself must be strong enough and stiff enough in both flexure and shear to resist the required earthquake forces without damage to the walls or any other part of the building. (2) The anchorage must be adequate to transmit loads normal to the wall from the wall to the diaphragm. (3) Connections must be adequate to transmit forces parallel to a wall from diaphragm to wall.

The floors of a bearing-wall building often have only a single layer of flooring laid at right angles to the joists. Such diaphragms are very weak and flexible in shear. If there is a layer of diagonal rough flooring below the finished flooring, the shear resistance of the floor will be much better. Whether it is adequate, depends on how the rough floor is laid and nailed. The joints in adjacent boards should be staggered at least two joist spaces, and there should be at least two 8 penny nails in each board at each bearing. The trouble is that the construction of the rough floor cannot be adequately checked without removing the finished flooring. Tests of full size wood floor panels reported in Reference 14 show the strength and deflection characteristics of wood floors.

The anchorage under item number 2 above is always inadequate if the walls are laid in lime mortar. The usual "T" type anchor, which was used in the collapsed walls of the building shown in Figure 30, is a poor anchor. It is usually inadequate even if the wall is laid in good cement mortar.

The wood floors in most bearing wall buildings do not meet the requirement described under item number 3. The end joists and the end blocking of joists are not bolted to the walls to transmit shear forces.

There is no practicable way to actually strengthen diaphragms having such deficiencies. New lateral bracing systems must be provided, making as few changes as possible in the existing floor structure. This can be done either with plywood and suitable connections, or by the use of steel bracing trusses.

New Plywood Diaphragm

It is not practicable to nail the new plywood directly to the existing floor joists. In the first place, all the existing joists may not be accurately spaced usually at 12 inches or 16 inches center to center, so that the joints in the 4-foot wide plywood sheets would not occur exactly over the center of a joist. If the plywood joint is not so centered, the edge distance of the nails will be reduced, and for a 2-inch nominal joist the tolerance in nail position for full nail strength is very small. Also it is difficult to exactly locate the existing joists for nailing purposes. They cannot be located from the position of the nails in the finished flooring, since this is always blind nailing.

The new plywood which is generally ½-inch thick for a plywood diaphragm,* is nailed to new 2″ × 4″ nailers placed at 24-inch centers on the existing floor, as shown in Figures 31 and 32. These 2″ × 4″'s must be fastened to the existing floor to prevent any possibility of upward buckling of the new diaphragm. Unless the existing floor is seriously out of

*For vertical live load on the floor, the ½-inch plywood spanning 24 inches is not adequate. A thicker finished floor would also be required. For a roof diaphragm however, ½-inch plywood is considered to be adequate for the smaller live load and it is generally used on spans up to 24 inches.

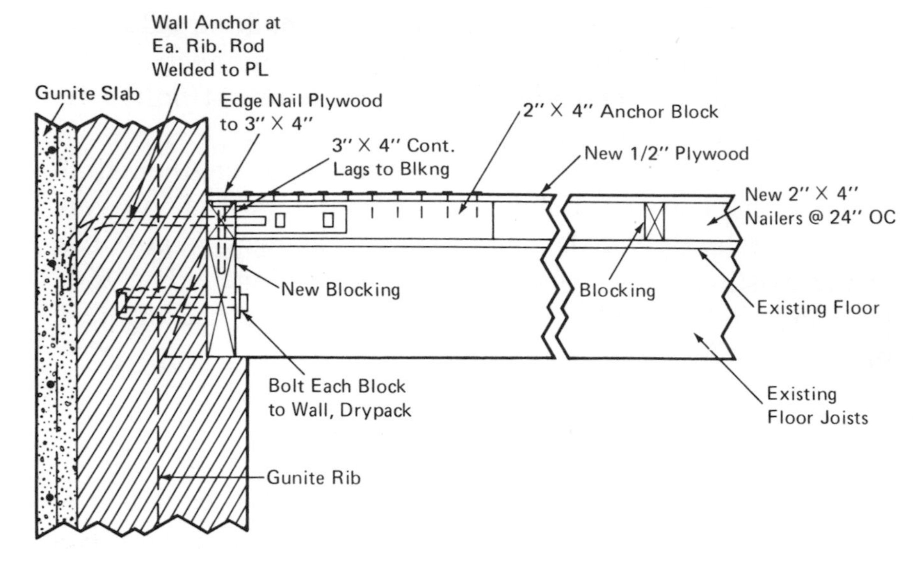

Figure 31. Exterior gunite wall strengthening and new plywood diaphragm.

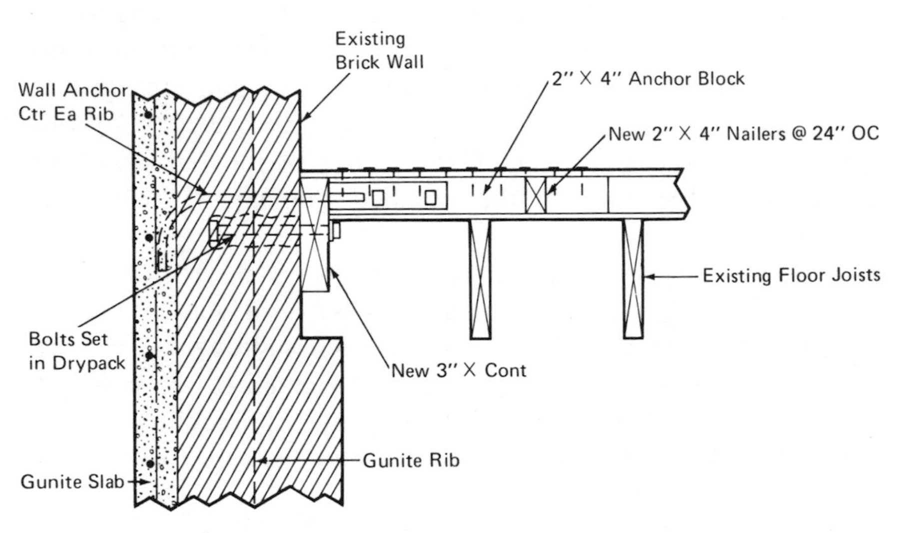

Figure 32. Exterior gunite wall strengthening and new plywood diaphragm, with floor joists parallel to the wall.

alignment, the buckling forces would be very small. They could be resisted either by toenailing to the existing floor or by small sheet metal clip angles nailed to the 2″ × 4″'s and screwed to the existing flooring. Such clip angles are available commercially.

The diaphragm constructed as shown in Figures 31 and 32 must be designed to meet the three requirements stated above.

A plywood floor diaphragm consists of 4 foot wide plywood sheets, nailed along the edges at opposite sides of each sheet to floor joists or in this case to 2″ × 4″ nailers, and along the other two opposite edges to blocking extending between the joists or nailers. Adjacent plywood sheets should meet accurately on the centers of the nailers or blocking, so as to give adequate edge distance for the nails. The sheets are also nailed at wider spacing, generally about 12 inches, to intermediate wood members to prevent buckling of the plywood sheets under load. This is termed a "blocked" diaphragm, and since each plywood sheet is connected to its neighbors by continuous nailing along all four edges, it has maximum shear strength. This same continuous nailing is used at the edges of plywood sheets at wood members along the boundaries of the diaphragm. The size and spacing of the nails at the edges of the plywood sheets determines the shear strength and stiffness of the diaphragm.

Comprehensive tests have been made of full size plywood diaphragms and shear walls, to determine their shear strengths and deflection characteristics. These tests were made and reported by the American Plywood Association and by the Oregon Forest Research Center. Current building codes, such as Title 21 and the Uniform Building Code, give diaphragm shear strengths for varying sizes and spacings of edge nails.[12] When the lateral earthquake load on the diaphragm has been determined, the maximum shear is calculated and the proper edge nailing is then selected from the applicable building code. Although 6 penny and 10 penny nails are occasionally used, by far the commonest size is 8 penny. It is not customary to change the nail spacing in a diaphragm because of changes in the shear; the spacing used is for the maximum shear.

Most building codes place a limitation on the maximum permissible span to width ratio of plywood diaphragms. This is to control lateral deflection. The span is commonly limited to four times the width of the diaphragm.

The plywood is designed to resist shear forces only. Flexural tension and compression in the diaphragm is resisted by continuous edge members or chords, to which the plywood is connected. For the diaphragm of Figures 31 and 32, the masonry wall itself forms the chord and the plywood is connected to it through wood members bolted to the wall. This bolted connection must be designed to resist the flexural horizontal shear in the diaphragm and also the end shear in the diaphragm at a shear wall.

The wall anchors are designed to resist the total earthquake load generated by the length of wall between anchors or ribs. The plywood should be nailed to the anchor block (see Figures 31 and 32) to develop the full strength of the anchor and the bolted connection to the anchor block.

The foregoing are the essential steps in the design of the plywood diaphragm shown in Figures 31 and 32.

Plywood Roof Diaphragm

The plywood roof diaphragm constructed as shown in Figures 32.1 and 32.2 must be designed to meet the same three requirements mentioned above for a floor diaphragm. However, the roof diaphragm is more apt to buckle under load. The two places that are particularly critical in this regard are

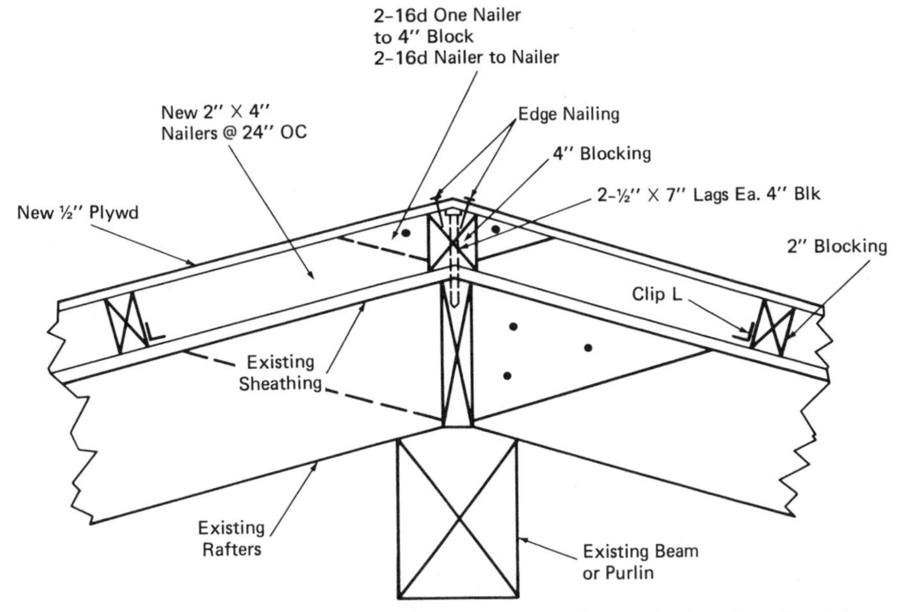

Figure 32.1. New plywood roof diaphragm at ridge. Similar detail at hip.

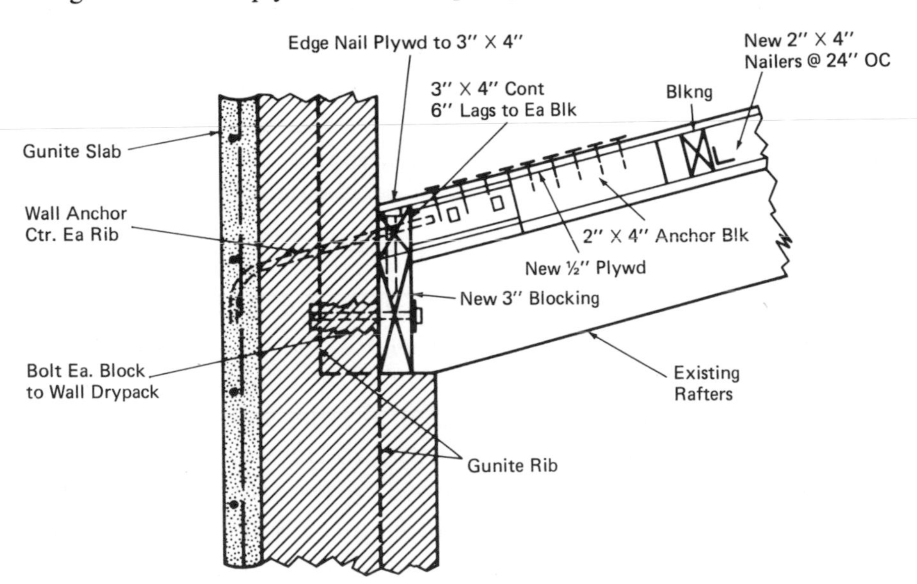

Figure 32.2. Plywood roof diaphragm at wall. Detail is similar to Figure 32.1 where rafters are parallel to wall.

the ridge and the hips, where there is an abrupt change in slope. At these points, 4-inch wide blocking between the 2″ × 4″ nailers is lagged to the existing rafters or blocking. The nailers lap here and one of them is nailed to the 4-inch block while the other is nailed to form the splice. By this means the new nailers are securely held down at the ridge and hips. At other portions of the roof there is not much more tendency to buckle than a floor. Use clip angles at 4′-0″ oc on the 2″ × 4″ nailers and one on each block. The horizontal leg of each angle should have 3-¾″ long screws of the largest diameter that will go through the regular holes in the angle. The vertical leg should have three nails to the 2″ × 4″ nailer.

On Detail 32.2 the 6-inch long lags through the 3″ × 4″ continuous and the bolts connecting the 3″ blocking to the wall should be designed to resist the shear between the diaphragm and the wall. This shear is produced either by a reaction force or it is horizontal shear in the diaphragm due to beam action, since the wall serves as the diaphragm chord. These remarks also apply to Figure 31.

The existing tar and gravel roofing should be removed before the new diaphragm is installed, and new roofing on the plywood will, of course, be required.

Steel Bracing Truss

The horizontal steel truss shown in Figure 33 performs the function of a diaphragm, in lieu of the deficient diaphragm action of the wood floor structure. The channel along the wall serves as a chord for the lateral truss and also acts as a beam to transmit the loads normal to the wall to the panel points of the truss. The inner chord of the truss is the existing wood floor beam, as shown in Figure 33. The beam must be spliced to resist tensile and compressive forces. The inner flange of the channel chord is tied at intervals to the wood floor above, to prevent buckling under the compression of

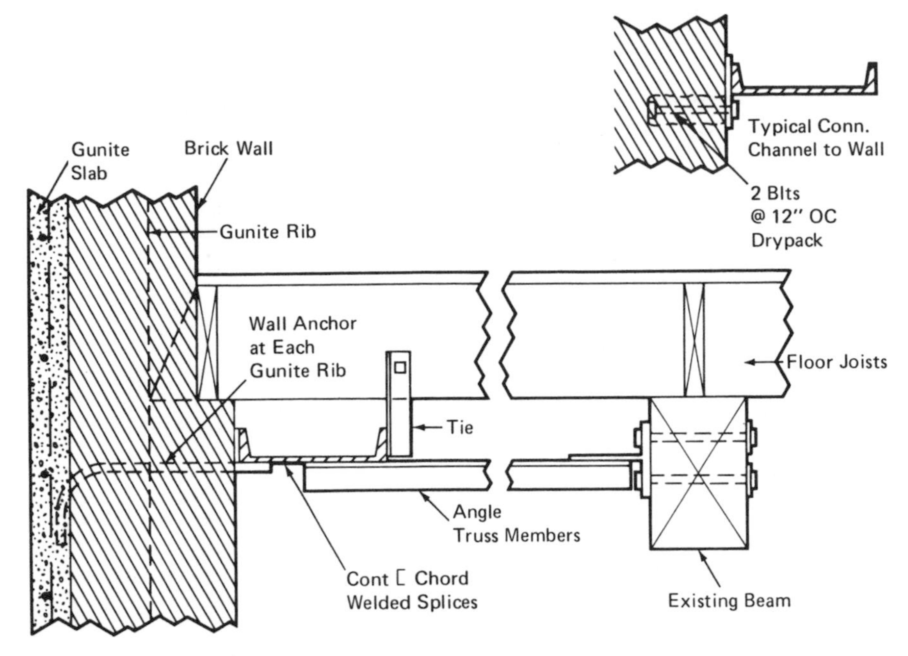

Figure 33. Exterior gunite wall strengthening and new lateral bracing truss.

chord and beam action. The bolted connections of the outer flange of the channel to the wall, are designed to transmit the reaction from the truss to the shear wall and to prevent buckling of the flange. The steel rods of the wall anchors are welded to the channel to transmit the lateral loads from the gunite ribs to the truss. In order to get maximum stiffness, it is best to weld all of the truss connections.

There is usually an elevator and stair passing through the truss. A warren-type truss should probably be used since this would give a maximum clearance for this purpose. Since the truss diagonals are rather long and may be in either tension or compression they should have lateral support. This can be accomplished by means of an angle extending from the center of the member to the channel chord.

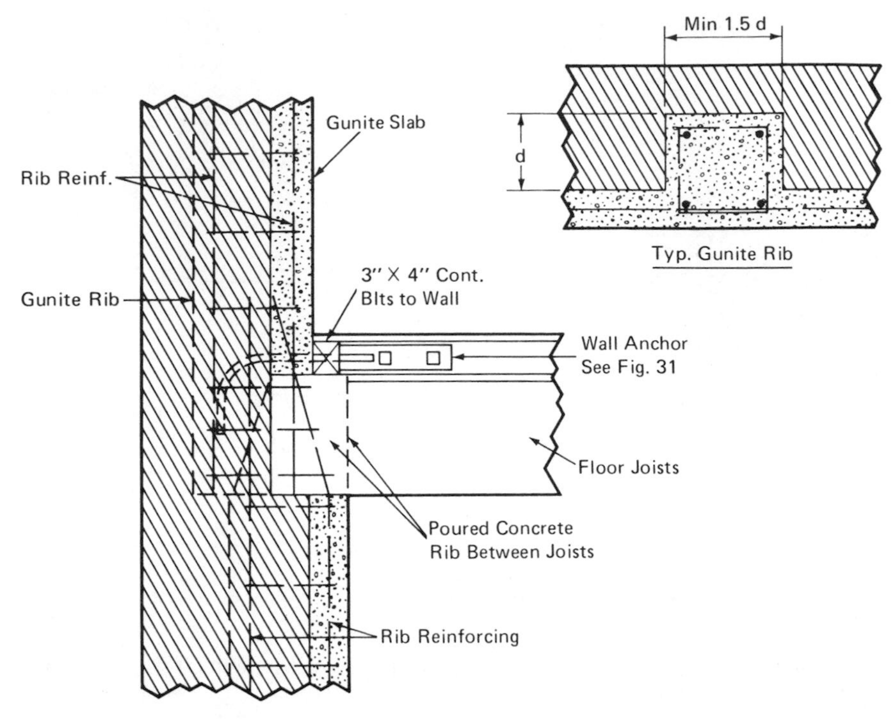

Figure 34. Interior gunite wall strengthening and new plywood diaphragm

Wood Diaphragm Versus Steel Truss

There are several considerations that will determine whether the wood diaphragm or the steel truss should be used in a given case.

Installation of the wood diaphragm would probably require the removal and replacement of partitions. Therefore, if the building has few partitions, or perhaps no partitions at all, as might be the case in a loft building, probably the wood diaphragm would be cheaper. Also the presence of ceilings would favor the use of plywood diaphragms, since a ceiling would probably have to be removed in order to install a steel bracing truss. The reinstallation of a suspended ceiling would be a large cost item.

There is also the matter of lateral deflection to consider. The diaphragm should be stiff enough to prevent cracking of the masonry walls during an earthquake. The maximum permissible lateral deflection of a masonry wall is largely a matter of engineering judgment. The lateral deflection is under better control with a steel truss than with a wood diaphragm. Most building codes limit the span to width ratio of a plywood diaphragm, as already mentioned, and if this ratio would be exceeded in a given case, steel truss bracing must be used.

USE OF PLYWOOD FOR SHEAR WALLS

Plywood is very useful for constructing shear walls, as well as diaphragms, for strengthening purposes. The plywood is applied and nailed in the same manner as for a floor diaphragm. The wall studs take the place of the floor joists and blocking between the studs serves the same purpose as blocking between the joists, that being to receive the edge nails at a horizontal butt joint in the plywood sheets. All shear walls are blocked construction to develop maximum shear strength and stiffness. The strength and stiffness of a shear wall can be approximately doubled by applying plywood to both sides of the studs, instead of to only one side. The same as for a diaphragm, the plywood on a shear wall is designed to resist shear forces only. Flexural tension and compression in the shear wall are resisted by end posts to which the plywood is nailed. The plywood is Douglas Fir exterior grade, usually ½-inch thick and with 8 penny nails.

Plywood shear walls are framed with 2″ × 4″ or 2″ × 6″ studs either at 16 inches or 24 inches on centers. The edge members or end posts are either 3-inch or 4-inch nominal thickness. This increased thickness of the post is to develop adequate bolt strength at the bottom connection to the foundation and also for bending strength, since the bottom connection is eccentric. This bottom connection is usually a clip angle with two bolts in the post and a single bolt to the foundation. The connection is designed to resist the uplift tension from overturning; post end bearing resists the compression. Building codes sometimes permit only a portion of the dead load on the post to be counted on for reducing the uplift, because of the vertical component of the seismic acceleration. Usually the clip angle of the post anchor is raised somewhat above the concrete of the foundation and the anchor bolt is tightened to take the initial slip out of the connection. This also increases the end distance of the lower bolt in the post. At the bottom of the shear wall the plywood is nailed to a plate or sill that is bolted to the foundation to resist the horizontal shear.

These plywood shear walls are frequently used for strengthening a wood frame building for resistance to earthquake forces. They are sometimes used in buildings having masonry or concrete walls but for this purpose the shear wall deflection is important; it should be quite stiff so as to give the masonry or concrete walls proper lateral support. The shear walls added to a building may be either interior or exterior. Generally such a shear wall would be of entirely new construction, because of the need for end posts and a specially bolted sill. Of course there must be some provision for getting the lateral load into the shear wall at the floors and roof. This would normally be by means of horizontal diaphragms, either new or existing. As has already been noted, such load distributing diaphragms are basic elements of any earthquake resistant structure.

When there are several plywood shear walls acting in series, such as those connected by a drag member along a wall, it is customary to distribute the lateral load to the shear walls in proportion to their widths. This implies that the walls all have the same deflection and that this deflection is primarily due to shear. The bending deflection of a shear wall is reduced to a minimum by tightening the anchor bolts as already described.

EXAMPLE OF PLYWOOD SHEAR WALL

Design a plywood shear wall in the exterior wall of a one story wood frame building in compliance with the Uniform Code. The wall is 2″ × 4″ studs spaced 16 inches on center, with a double top plate supporting 2 inch by 12 inch rafters spanning 25 feet 0 inches to an interior support. The shear wall is 12 feet 0 inches long and 8 feet 0 inches high. It will resist a 7000 pound seismic load. The roof, with a plywood top deck, tar and gravel roofing and plaster ceiling, weighs 18 pounds per square foot. The wall is plaster on each side and weighs 22 pounds per square foot.

Total dead load on each end post = $(18 \times 15.5 + 22 \times 8)$ $6.67 = 3000$ pounds.

Horizontal shear = $7000/12 = 585$ pounds per linear foot.

Use ½-inch plywood. Note that the plywood must be let into the studs ½ inch in order that the plaster on the plywood will be flush with the plaster elsewhere on the wall. Then this shear requires 8 penny nails at 2 inches on center at the edges of all plywood sheets. This gives a shear resistance of 600 pounds per linear foot for a blocked diaphragm. It will be noted on Figure 34.1 that there is a line of blocking centered on the horizontal joint in the plywood at mid-height of the shear wall, to receive edge nails. In addition to edge nails at 2 inches on center at the perimeters of all plywood sheets, there should be nails at a maximum of 12 inches on center in the interior of the sheets. The purpose of these nails is to eliminate any chance of buckling of the plywood sheets under the action of shear stress.

The seismic load on each end post = $7000 \times 8/12 = 4700$ pounds.

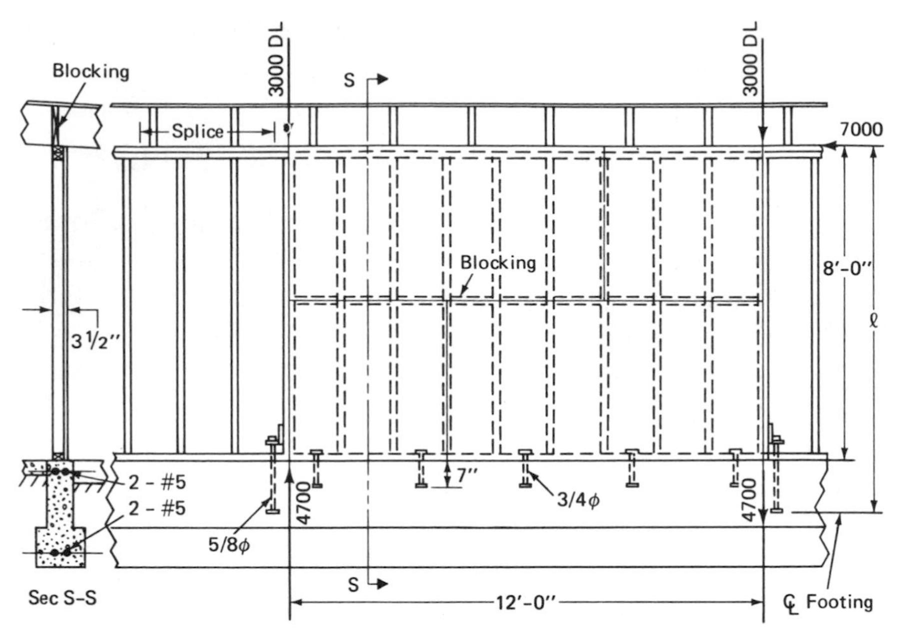

Figure 34.1. Plywood shear wall in the exterior wall of a one story building.

Use 3″ × 4″ end posts. It is assumed that there is a 0.25g vertical acceleration for the following.

Net uplift on an end post is 4700 − 3000 × 0.75 = 2450 pounds.

For end post anchorage to the foundation use a 6″ × 4″ × ⅝″ angle, which should be 3 inches long since the end post has been cut to about 3 inches. Use 2-⅝-inch bolts to the post, and 1-⅝-inch anchor bolt to the foundation. A ⅝-inch bolt is good for 1000 pounds in single shear in Douglas Fir members and this may be increased 25 percent if one member is a steel plate. The bolt value may be increased 1/3 for a seismic load, then

2-⅝ inch bolts = 2 × 1000 × 1.25 × 1.33 = 3300 pounds.

2-½ inch bolts = 2 × 650 × 1.25 × 1.33 = 2150 pounds, less than 2450, not enough.

The bottom bolt in the post should be not less than 7 bolt diameters or 4⅜ inches from the end of the post. The two bolts should be spaced not less than 4 bolt diameters or 2½ inches apart. The gauge of the anchor bolt in the outstanding leg of the 6″ × 4″ angle should be as small as possible and still provide room to turn the nut without contacting the fillet of the angle. This reduces bending on the angle. Suppose the gauge is 2 inches. Then the bending moment in the angle is:

$$2450 \times (2 - 0.31) = 4150 \text{ inch pounds.}$$

The moment of inertia of the 3-inch wide angle is:

$$I = \frac{3 \times 0.625^3}{12} = 0.061$$

and the bending stress in the angle is:

$$S = \frac{4150 \times 0.31}{.061} = 21{,}000 \text{ psi}$$

The anchor bolt should have a net area at the root of the thread of:

$$A = \frac{2.45}{20 \times 1.33} = 0.09 \text{ square inch}$$

A ⅝-inch bolt provides an excess area.
The section area and moment of inertia of the end post are:

$$A = 3 \times 2.5 = 7.5 \text{ square inch}$$

$$I = \frac{3 \times 2.5^3}{12} = 3.9$$

The direct tension plus bending stress in the end post is:

$$S = \frac{2450}{7.5} + \frac{2450 \times 1.25}{3.9} = 790 \text{ psi}$$

The foundation sill cannot be continuous under the end posts because the cross grain compression on the sill would be much too high. Therefore extend the end posts through the sill, so that they bear directly on the concrete. Then the compression on the concrete or the end grain bearing on the post is not excessive.

Use ¾-inch bolts to connect the redwood sill to the foundation. The shear value of a ¾-inch bolt in 3000 pound concrete is $1780 \times 1.33 = 2200$ pounds and in the redwood sill it will develop $1160 \times 1.33 = 1520$ pounds then:

$$\text{Number of sill bolts} = 7000/1520 = 4.5\text{; use 5.}$$

The double top plate must perform two functions. It must serve as a drag member to deliver the lateral load to the several shear walls in the wall. It must also serve as a chord member for the plywood roof diaphragm, when it resists lateral loads normal to the wall. The splice in such a double plate is generally nailed using 16 or 20 penny nails. The 16 penny nail is good for $107 \times 1.33 = 142$ pounds in single shear, and the 20 penny nail will resist $139 \times 1.33 = 185$ pounds. Occasionally a bolted splice is used if the number of nails requires an excessive length of splice.

There is continuous blocking between the rafters where they rest on the stud wall, as shown on Sec. S-S. This blocking serves to transmit lateral load parallel to the wall, from the plywood roof deck to the double plate. For this purpose toenails are used in the blocking. A toenail is allowed $2/3$ of the shear value of a nail driven at right angles to the face of a member. Toenails are often 10 penny which is good for $94 \times .67 \times 1.33 = 84$ pounds. If four toenails are used for each side of the blocking, a shear value of $84 \times 4 = 340$ pounds per linear foot of plate can be developed. This would generally be adequate for either the horizontal shear between the roof diaphragm and the double plate chord, or for transmitting load parallel to the wall, from the roof diaphragm to the shear walls.

There is no way to accurately calculate the stress in the continuous footing supporting the shear wall, due to the overturning moment produced by the lateral force. A conservative assumption is that this moment is resisted by bending on two sections of the footing, one section each side of the shear wall.

To resist this bending there should be reinforcing top and bottom of the footing, as shown on Sec. S-S. On the detail of Figure 34.1, the top bars would be in tension on the right hand side of the shear wall and the bottom bars would be in tension on the left side.

The overturning moment is the lateral force of 7000 pounds times the distance 1 (in inches) from the line of action of this force to the center line of the footing, as shown on Figure 34.1. Then, using the conventional nomenclature for reinforced concrete, this method of calculation can be expressed as follows:

$$2A_5 f_s jd = 7000 \times 112$$

If it is assumed that d is 30 inches, and using $f_s = 26{,}600$ psi, then:

$$A_5 = \frac{7000 \times 112 \times 0.5}{26600 \times .875 \times 30} = 0.56 \text{ square inch}$$

This steel area would require two number 5 bars. Since this reinforcing is only required at the shear wall, one number 5 bar need only extend say 8 feet 0 inches beyond the post each side, which would require a bar 28 feet 0 inches long. One number 5 bar should be continuous, since no continuous footing should be unreinforced.

DESCRIPTION OF A STRENGTHENING PROJECT

The Author was the structural engineer on an earthquake strengthening project that will illustrate the use of a plywood diaphragm. The building to be strengthened is a large one story school with concrete exterior walls, wood joist floors raised above grade and a wood roof. The typical plan is 25-foot wide classrooms between exterior walls, with outside corridors. The plaster ceiling on 2″ × 4″ joists was hung from trussed rafters. In addition to the deficient earthquake resistance of the building, these rafters were not strong enough to meet code requirements for dead and live load.

The first operation was to remove the suspended ceiling. The rafters were not disturbed. Removal of the ceiling reduced the load enough so that the rafters would be stable during the reconstruction. Next, continuous steel ledger angles were bolted to the concrete walls at the ceiling level, using expansion bolts to give both shear and tensile resistance without penetrating the wall. Wood joists were then installed, spanning across the class rooms and supported on the ledger angle at each end. Plywood was nailed on the joists and to blocking between the joists, to form the diaphragm. The rafters were given permanent support near their centers, on two lines of studding resting on the plywood. Alternate joists were lagged to the ledger angles to transmit seismic loads normal to the walls, and to the diaphragm. Seismic loads parallel to the wall were transmitted from the diaphragm to the wall by nailing the plywood to end blocking between the joists, each of these blocks being lagged to the ledger angle. The concrete walls themselves formed the edge members of the diaphragm for resisting flexural stresses of tension and compression.

The new plywood diaphragm would have been too long without any lateral support between the two end reactions at exterior walls. The maximum permissible length to width ratio of a wood diaphragm is limited by code, to prevent excessive lateral deflection.[12] This is particularly important where the diaphragm affords lateral support to concrete walls. Accordingly at about the center of the length of the diaphragm it was connected to a new concrete shear wall. This was at the location of a partition between two class rooms.

An interesting feature of this project was the X-ray method that was used for investigating the nature of the steel reinforcing in the concrete walls. The building was old and although there was some reference in the specifications to concrete reinforcing bars, there was no information as to their location, size or spacing. A radioactive isotope in a heavily shielded container, was moved around the building on a small hand truck. Radiation from the isotope was directed against the wall and recorded on a large photographic plate on the opposite side. Although some of these walls were as much as 18 inches thick, the bars showed quite clearly as dark shadows on the photograph. However, the shadow was not sharp enough to indicate the size of the bar. This was easily determined when its position was known, by cutting into the wall and exposing the bar. The concern that did the X-raying developed and printed the plates right on the job, so that it could immediately be determined if satisfactory information was being obtained.

HIGH RISE CONCRETE

In Caracas, Venezuela, after the 1967 earthquake, there was a great deal of strengthening and repair of high rise buildings. In general this work was well designed by competent structural engineers and in accordance with the same procedures used in the United States. Usually the buildings had reinforced concrete frames that were designed to support the vertical loads and also resist the lateral forces. In Caracas, a large number of these buildings were high rise.

The strengthening of the Mene Grande Building in Caracas is a good example of such a project and it is of particular interest because of the great increase in strength. This 16 story and basement building was complete in 1966 and was designed at that time in compliance with the latest available standards, for a base shear of 1.43 percent of gravity. The design of the strengthening provides for a 3.0 percent base shear, more than double the original load.

This building had the usual concrete frame that was designed to resist both vertical and lateral forces. After the earthquake, most beams in the building had cracks near the column up to about the sixth floor, decreasing above that level. The most important damage, however, was the failure in the first story of seven of the eight corner columns. These columns were A2, A4, A6, F2, F4, F6 and F8, as shown on the typical floor framing plan, Figure 35. The cause of these failures was probably the increased overturning moment due to the stiffening effect of the solid tile end walls. This effect of infilled masonry walls was discussed in Chapter 5.

Figure 35 is a plan of the strengthening. This consists of four new concrete shear walls at the ends of the wings, which were designed to resist all the north to south earthquake forces, and eight new bracing bents that resist the east to west forces. Figure 36 shows the typical reinforcing at a new corner column, which consists of the column reinforcing, the reinforcing at the end of the new concrete shear wall and the reinforcing of the new bracing bent beam where it frames into the corner column. Figure 37 is the typical reinforcing at the opposite end of a new bracing bent, including the added reinforcing at the existing column and the reinforcing of the new beam framing into the column. All the new corner columns having the reinforcing shown in Figure

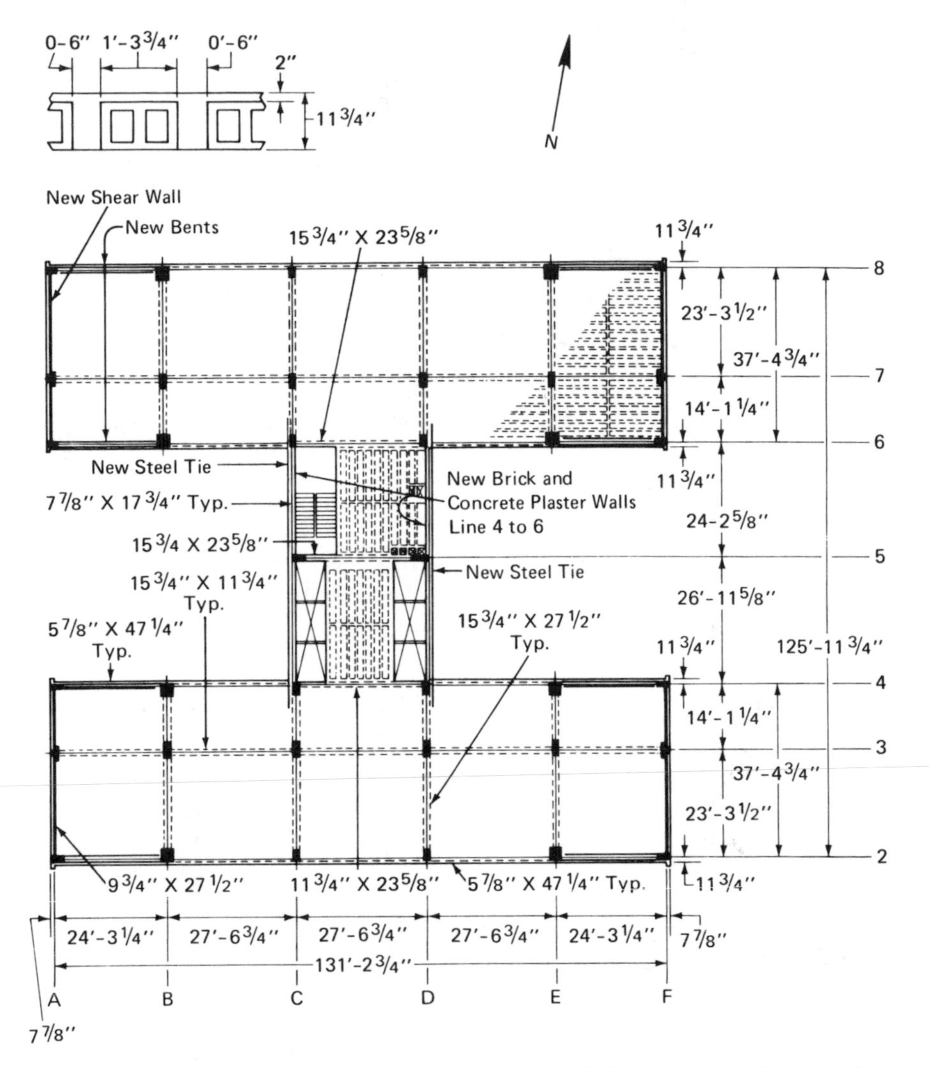

Figure 35. Typical floor plan Mene Grande Building, Caracas, Venezuela, showing earthquake strengthening.

36 and the reinforced existing columns as shown in Figure 37 penetrate all floors and are continuous from foundation to roof. It will be noted from Figure 36 that the new beam of the bracing bent is above the floor. This is convenient for

Figure 36. New corner column at the end of a new shear wall and the end of a new beam of a bracing bent, Mene Grande Building. Photograph was taken during the strengthening of the building.

Figure 37. End of new bracing bent, showing reinforced column and reinforcing of new beam above the floor, Mene Grande Building.

placing reinforcing and pouring concrete, and it involves a minimum of cutting of existing concrete. In Figure 37 it is seen that there is no reinforcing on the near side of the existing column except some ties, so that the floor girder framing into this side of the column was not disturbed. Also note that the existing column was roughened to improve the bond between the new concrete and the existing concrete. The new shear walls, the reinforcing of one of which is shown penetrating the near side of the column in Figure 36, were placed inside the existing solid hollow tile exterior end walls.

MULTISTORY STEEL FRAME

Very little experience or thought has been given to the strengthening of the moment resisting steel frames of multistory buildings for resistance to earthquake forces. This is probably because there are relatively very few multistory steel frame buildings in areas where severe earthquakes have occurred. Nearly all the buildings in these areas have been reinforced concrete. For example, in Caracas, Venezuela, in 1967, there were only two high rise buildings of structural steel; all the rest were concrete. In Anchorage, Alaska, at the time of the 1964 earthquake, there were only two high rise buildings and both of them were concrete. During the San Fernando earthquake, all the multistory buildings in areas of severe shaking were concrete.

Apparently the only practicable way to strengthen a moment resisting steel frame is by adding braces, either "X" bracing if the wall has no windows, or "K" bracing or corner bracing if there are openings. In many cases the columns would probably be adequate for axial stress due to overturning, since they would be entirely or largely relieved of bending stress. However, the braced bent would be stiffer than the original moment resisting frame and would attract more seismic load. This would increase the overturning moment and axial stress in the columns.*

If material must be added to a column to increase its axial load capacity, the resulting section would be very unequally stressed, since the original material must carry the full dead load. It would seldom be practicable to relieve a column of dead load before it is strengthened.

ADMINISTRATION AND COORDINATION OF STRENGTHENING PROJECTS

An earthquake strengthening project requires not only structural plans and specifications but also plans and specifications for architectural work and for mechanical and electrical work, since they are all usually affected by the strengthening work. The coordination of the structural work with that of the architect and the mechanical and electrical engineers is very important. If this coordination is not done well, there may be serious difficulties during construction and extra costs to the owner. The coordination is done by either the structural engineer or the architect, depending upon which of them is the prime contractor for the preparation of plans and specifications. The prime contractor deals directly with the owner of the building and employs the other professionals. The Author has worked both ways, either as prime contractor, or as a consulting engineer employed by the architect. It is better that the structural engineer be the prime contractor, since the most important work is structural and all other work should conform to the structural requirements.

On the project for strengthening a school building, pre-

*It should be noted that this method of strengthening a moment resisting frame will reduce its ductility.

viously described, there was a serious lack of coordination between the structural work and the electrical work. This illustrates very well the importance of good coordination.

The electrical contractor on this project did not know that all the existing ceilings in the building would be removed and reinstalled as part of the strengthening work. All the conduits for the electric lights were supported in the attic on the original 2″ × 4″ ceiling joists. It would, therefore, be necessary for the electrical contractor to remove and reinstall all these electrical circuits. For this work, which he did not figure on, the electrical contractor requested an extra of $20,000 on his contract. Although this extra cost was a relatively small percentage of the entire cost of the strengthening project, it just about doubled the cost of the electrical work.

This strengthening project was done for the Los Angeles City School Department, and they have a standard requirement that any requests for extra costs, beyond a certain amount, on strengthening projects must be referred to a Board of Arbitration for review and settlement. This Board was comprised of three members, one structural engineer acceptable to both parties, a member appointed by the School Department, and a member appointed by the contractor involved. In this case the Board decided in favor of the electrical contractor for the total amount of $20,000. The Author considers this to be a very inequitable decision.

The specifications for the project stated that the complete plans were available at the Board of Education, for inspection by any subcontractor. It is impossible to show, on the plans for each subcontract, the strengthening work that will be done. Any subcontractor is negligent if he figures an earthquake strengthening job without looking at the complete plans to see what portions of the structural work may affect his work.

PREVENTION OF NONSTRUCTURAL DAMAGE

Nonstructural earthquake damage is generally caused by excessive lateral movement of the building. This was illustrated by the great difference in the nonstructural earthquake damage to two adjacent multistory buildings in Managua, one stiff and one flexible building, as described in Chapter 5, with regard to shear walls.

All equipment or furnishings that are hung from the ceiling such as light fixtures, ducts, piping and heating units, should be braced so they cannot swing. Long pendant mounted fluorescent lighting fixtures that are free to swing, usually break loose.

Equipment that is not securely fastened to the floor such as boilers, furnaces, water heaters, storage tanks and air conditioning units will be displaced or overturned. Generally utility lines to these units will be broken. Vibration isolation supports are particularly vulnerable to earthquake forces. Storage cabinets and storage racks should be securely anchored to walls or partitions.

Partial height partitions should be stable, particularly those at toilet stalls which are often of marble.

Often the greatest single item of expense concerning nonstructural damage, is for the repair of cracked plaster in partitions. During an earthquake each story of a building undergoes a shear distortion, which is a horizontal movement of the upper floor of the story with respect to the lower floor. If a partition in the story is connected to the structure so that it is forced to undergo this same shear distortion, and if this is great enough, the partition will be "X" cracked. This cracking can be prevented if the partition is mounted so that it is free to slide at the ceiling and has a clearance where it abuts either an exterior wall or a column. This type of mounting has been used in new buildings at little additional expense, for example, the 43 story La Latino Americana Building in Mexico City. It is not very easy to cut loose partitions in an existing building so as to accomplish the same result. Another solution, of course, is to stiffen up the building, probably by the use of shear walls, so that the shear distortion in a story is not enough to crack the plaster in the partitions. This might be done for a new building, but it is not practical for an existing building, unless it is simply part of an overall strengthening project.

Another result of shear distortion of partitions is the racking of door frames, so that doors are jambed shut or will not close.

It is shear distortion of a story that breaks the glass in windows that are rigidly connected to the structure of the building. A window should have clearance around it to permit relative movement in its plane of about ¼ to ½ inch. This is also a feature of the La Latino Americana Building.

Ceilings sometimes fall during an earthquake but this does not result in as serious damage or pose as much of a hazard as it used to, because ceilings are now generally acoustical tile, not plaster. Tiles are mounted on furring channels or "T" members, to which they should be securely fastened. Some clip type fasteners break loose during the violent motion of an earthquake. The furring should be securely fastened laterally at walls and partitions.

Elevator damage often results from violent movement of the counter weights. These should be retained in some way in their guides.

So far as exterior damage is concerned, the important thing is to provide adequate vertical supports and secure anchorage for all ornamentation, screen walls, and wall panels. Prefabricated wall units are now commonly used, particularly on high-rise buildings. These are often precast concrete and they are heavy. Good building codes have detailed requirements for their vertical support and for anchorage to the structure. They should be mounted so as to allow movement relative to the building structure, in the plane of the panel, without damaging the panels or breaking them loose.

Brick veneer sometimes breaks loose and falls, particularly if it is on wood construction. The veneer is rigid but there may be movement in the wood backing due to shrinkage and nail slip. Since veneer often occurs as an architectural feature over an entrance, its falling may be particularly hazardous. Title 21 has a very good specification for the vertical support and anchorage of brick veneer on wood.[12]

Parapet Walls

A very common type of non-structural damage involves the throwing down of unreinforced masonry parapet walls. If the wall faces a street front it constitutes a serious hazard since it will fall on the sidewalk. After the Long Beach, California, earthquake in 1933, the Author saw debris from fallen parapets on most of the sidewalks in the business area.

City ordinances often require the removal or strengthening of parapet walls even if there are no other building strengthening requirements. For example, the City and County of Los Angeles has a parapet ordinance and, to comply with this ordinance, about eight thousand buildings have had such parapet corrections made. The excellent work carried out under this ordinance saved many persons from injury or death due to the San Fernando earthquake.

Building codes usually require a high seismic coefficient for parapet walls, since experience has shown that they are particularly liable to earthquake damage. For example, the SEAOC Seismic Code requires that a cantilever parapet wall in an area of high seismicity be designed to resist a lateral force equal to its own weight.

The usual method of strengthening is to bolt a continuous angle to the top of the wall and to give this angle lateral support at intervals by means of a diagonal brace extending down to the roof. The connection of the brace to the roof must resist either tension or compression and it must be waterproof.

7

Repair of Earthquake Damage

After an earthquake, repair of the damage to a building is essential, so as to restore it to a condition that is usable and safe to the occupants. In this regard, it has sometimes been considered adequate to simply restore the building to its original condition. However if this is done and the building undergoes another earthquake of about the same intensity, it might be expected that there would be a repetition of the same damage. This is exactly what happened to the Liwayway Hotel in Manila.[10] This is a nine story reinforced concrete building built in 1930. In 1968 this building was severely damaged by an earthquake. This damage was repaired by removing all loose concrete, welding in new reinforcing bars where necessary and pouring new concrete. Smaller cracks were opened up by chiseling and then dry packed solid. The repairs were intended to restore the building as nearly as possible to its original condition. In 1970, 20 months later, Manila was subjected to another earthquake that was very similar to the one in 1968. The result was damage to this building that was nearly identical with the damage that it sustained in 1968.

This experience with the Liwayway Hotel demonstrated that a concrete building that is damaged by an earthquake can probably be restored to its original condition, at least so far as earthquake resistance is concerned. Particularly since better methods of repair are now used, for example, the use of epoxy. This may not be true of a building having a steel moment resisting frame.

The money spent in 1968 for repair of the Liwayway Hotel was a very poor investment. However it would normally be expected that there would be many years, instead of only 20 months, between two high intensity earthquakes in the same area. There are two objections to doing damage repair only. There is the possibility that a future earthquake might exceed in intensity the one for which repairs were made. Also there might be a short time interval between the first and future earthquakes, as happened in Manila.

Even though a damaged building will be strengthened, it is advisable to first make such repairs as are necessary to restore it as nearly as possible to its original condition. The strengthening job then starts with a structure whose earthquake resistant properties are known.

USE OF EPOXY

Epoxy adhesives are now widely used for repairing cracks in concrete walls, columns, slabs, and beams. The method consists of pumping or injecting low viscosity epoxy grouting materials under pressure into cracks and voids.

A crack must be prepared to receive and retain the grout. It is first cleaned by vacuuming or blowing out with compressed air. For retention of grout, the crack is "V'd" out at the surface and the "V" is sealed by filling with epoxy paste. To receive the grout small holes or "ports" are drilled over the center of the crack at intervals of 6 inches to 3 feet, depending on the length of the crack. At each port some type of fitting is installed through which the grout is injected from a nozzle at a pressure of usually 20 to 40 pounds per square inch, from a metering mixing pump. A popular fitting now available is a polyethylene one-way valve. Injection is begun at the lowest port until epoxy emerges at the next port. Then injection is shifted to this port and it proceeds from port to port. All ports must be sealed.

It has been found that injecting epoxy in this manner generally gives good penetration. Core samples are taken to check the penetration. Most specifications require the epoxy to penetrate at least 90 percent of the crack volume in the core. If core tests fail, additional injection work may be required

Figure 38. Epoxy repair of earthquake damaged concrete shear wall at the Indian Hills Medical Center.

but this is not always effective. It is better to make every effort to secure good initial penetration, followed by continuous inspection of the operation by well qualified personnel.

The bond of epoxy with concrete is tested either by splitting tests of concrete cylinders that have been repaired with epoxy, or by tests of concrete beams that have been repaired. During tests, the specimen should break through the concrete, not through the epoxy joint where it has been repaired.

Epoxy should be a two part material that meets standard specification requirements for flexural strength, tensile strength, compressive strength, and elongation.

Figure 38 shows epoxy repair work on a concrete shear wall of the Indian Hills Medical Center that was damaged by the San Fernando earthquake (see Figure 15). The epoxy sealing of the cracks has made them easily visible. It is seen that there is a network of closely spaced diagonal cracks throughout the entire height of the wall. The wall was evidently greatly overstressed in shear. Later the wall was strengthened by applying an 8-inch thickness of gunite with vertical and horizontal bars. This is an excellent procedure, first repair then strengthen.

CYCLIC LOADING TESTS OF EPOXY REPAIRED CONCRETE BEAMS

At the University of California in Berkeley,[20] severely cracked reinforced concrete cantilever beams that had been repaired with epoxy injection in the manner described above were tested under cyclic loading to simulate the periodically reversed load sustained by the members of a moment—resisting concrete frame during an earthquake.

The repaired beams showed excellent energy absorption and were found to be capable of resisting numerous applications of cyclic loads. They were stronger than the un-

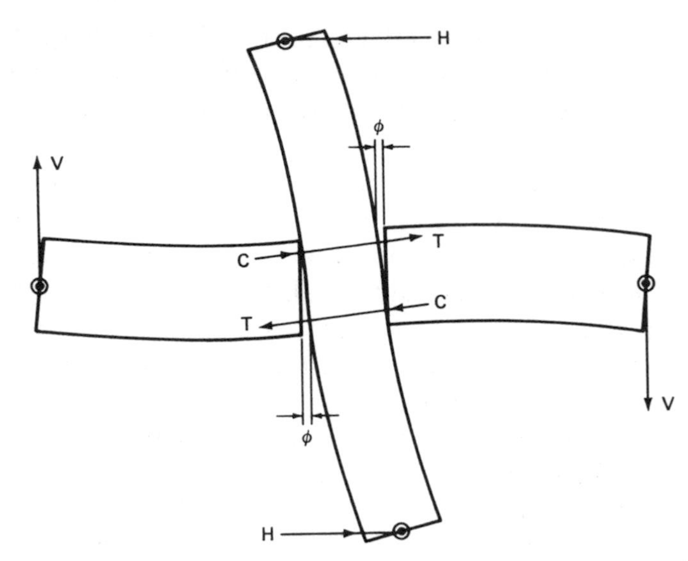

Figure 37.1. A typical interior joint of a concrete moment-resisting frame subjected to a lateral load.

cracked beams due to strain hardening of the reinforcing steel but slightly more flexible, probably due to incomplete filling of the cracks with epoxy. It was concluded that epoxy injection is a very effective method of repair even for members subjected to cyclic loading, except for one limitation, as illustrated by Figure 37.1. As shown in Figure 37.1, the total force $T + C$ due to lateral load tends to slide the beam reinforcing through the column, thereby producing bond stress. Obviously, this bond stress will not be produced by vertical loads since for this type of loading the forces T or C each side of the column will generally balance. If there is bond failure, the bond cannot be completely restored by epoxy injection, this is the limitation.

BOND TESTS

A series of five cyclic bond tests stimulating the condition shown in Figure 37.1 was carried out on #6, #8, and #10 bars. These bars were imbedded in 4000 psi normal weight concrete blocks that were reinforced like the columns of a ductile concrete frame. A compressive force was applied to one end of the test bar and a simultaneous tensile force of the same magnitude to the opposite end. The two forces were reversed periodically to give cyclic loading. Cyclic loading produced lower bond strengths than did loads in only one direction. For the five tests, the bond stress at total bond failure ranged between 1580 and 1904 psi. It was found that bond strength is strongly dependent on concrete compressive strength. Lightweight concrete of comparable strength gave lower bond strength.

Analysis of the Bond Stresses

Reports of these tests (Refs. 20 and 21) do not give any idea when, in actual practice, bond failure is likely to occur in a moment resisting concrete frame subjected to earthquake loads (EL). In these reports it simply states that for severe cyclic loading of such a frame, severe bond degradation can take place. The question is, what constitutes severe cycling loading and is it likely to occur? This is important, since bond failure cannot be repaired by epoxy injection. The following simple analysis provides a reasonable answer to this question.

For a beam of the frame

EL stress = (DL + LL) stress × 1.5

Reinforcing = #8 bars, area 0.79 square inch perimeter 3.14 inches

Steel intermediate grade (DL + LL) stress = 20,000 psi

Concrete 4000 psi, $n = 8.0$, (DL + LL) stress = 1800 psi

For these parameters the earthquake forces are

$$T = 20{,}000 \times 1.5 \times 0.79 = 23{,}700 \text{ pounds}$$
$$C = 1800 \times 1.5 \times 2 \times 8.0 \times 0.79 = 34{,}100 \text{ pounds}$$

In the calculation of C, the compressive stress in the bar is assumed to be $2n$ times the compressive stress in the concrete at the same point. The use of $2n$ rather than n is a common assumption and is intended to allow for the effect of creep and shrinkage of the concrete. It probably gives excessive bar stress.

For a column width of 16 inches

$$\text{Bond stress} = \frac{23{,}700 + 34{,}100}{3.14 \times 16} = 1150 \text{ psi}$$

This bond stress is 73% of the lowest bond stress (1580) developed in these tests. Evidently bond failure is not likely to occur and the method of repair by epoxy injection should usually be fully effective.

Effect of Bar Slip on Joint Yielding

In the bond tests described above there was considerable bar slip. This will produce a yielding of the interior joints of a frame. This yielding is simply a rotation of the end of the beam with respect to the column. This rotation is indicated by the angle ϕ in Figure 37.1. The usual assumption is that the joint is rigid so that the beam and column rotate through the same angle, θ, as indicated on Figure 1. No method has yet been developed for calculating the angle ϕ and its effect on the deflection of the frame. A complication is the fact that the yielding of the interior joints of a frame would be much greater than the yielding of the joints at an end column since the latter involves only T and C and not $T + C$. Also the bars would be hooked at an end column unless the beam is continuous through the column.

SPALLED CONCRETE

Where concrete has broken out, any loose concrete is removed and the spalled area may be filled with nonshrink grout or dry pack. Before placing this material, a liquid epoxy adhesive may be applied to the surface of the spalled area to improve the bond. As an alternate, the depressed spalled area may be filled with epoxy concrete or mortar. This is simply a liquid two component epoxy, mixed with sand and aggregate.

BUCKLED REINFORCING

Where reinforcing has buckled, the bent portions of the bars are cut out and replaced with new bars of the same size, with butt welds to the original bars. Ties or stirrups that are broken should be repaired by welding.

STRUCTURAL STEEL FRAME

Any steel members or connections that have been stressed beyond the yield point must be replaced, otherwise the ductility of the structure for resistance to future earthquakes, will be reduced. These critical regions which may develop inelastic deformations are localized and a large part of the structure may be free of such damage. The critical points in a moment resisting steel frame are likely to be the ends of beams or girders and their connections to columns. The portion of a column extending from just above a beam or girder to just below it, may be critical; although a good design avoids the chance of yielding in the columns because of its possible effect on the stability of the structure. Inelastic deformations in the "joint" region of a column may significantly add to the joint rotation and the deflection of the frame. That is, it adds to the ductility of the structure.

Damaged steel members or portions of members can be cut out and replaced by welding. Welding procedures should be carefully planned to prevent the development of residual stresses due to the welding operation. This is particularly important where thick material is involved.

Determination of the extent of damage may require the removal of ceilings, plaster or sprayed on fireproofing, and other finishes, at least in suspected regions, which may be indicated by cracks or other visible damage. A detailed review of the design of the structure should also be made, to help in locating critical areas.

References

1. The San Fernando, California Earthquake of February 9, 1971. Geological Survey Professional Paper 733.
2. Green, Norman B. Some Structural Effects of the San Fernando Earthquake. *J. Am. Concr. Inst.*, May 1972.
3. Suyehiro, Kyoji. Engineering Seismology—Notes on California Lectures. *Proc. Am. Soc. Civ. Eng.*, **58**, No. 4, May 1932.
4. Lateral Forces of Earthquake and Wind. Joint Commission of the American Society of Civil Engineers and Structural Engineers Association of Northern California, *Trans. Am. Soc. Civil Engrs.*, Vol. 117, 1952.
5. Hanson, Robert D. and Henry J. Degenkolb. The Venezuela Earthquake July 29, 1967. Am Iron Steel Inst.
6. Jennings, Paul C. The Effect of Local Site Conditions on Recorded Strong Earthquake Motions. *Proc. Annual Convention Structural Eng. Assoc. California*, 1973.
7. Jephcott, D. K. and Hudson, D. E. The Performance of Public School Plants during the San Fernando Earthquake. California Institute of Technology, September 1974.
8. Berg, Glen V. and Stratta, James L. Anchorage and the Alaska Earthquake of Ma.ch 27, 1964. American Iron and Steel Institute.
9. Jennings, Paul C. (Ed.), Engineering Features of the San Fernando Earthquake, February 9, 1971. California Institute of Technology.
10. Dean, R. Gordon. Observations of Earthquake Damage in Manila—1970 Earthquake. *Proc. Annual Convention Structural Eng. Assoc. California*, 1970.
11. Fintel, Mark. Ductile Shear Walls in Earthquake Resistant Multistory Buildings. *J. Am. Concr. Inst.*, June 1974.

12. Title 21 of the California Administrative Code. Office of Architecture and Construction.
13. Sharpe, Roland L., Kost, Garrison, and Lord, James. Behavior of Structural Systems Under Dynamic Loads. National Bureau of Standards, Building Science Series 46, February 1973.
14. Green, N. B. and Horner, A. C. Earthquake Resistance to Timber Floors. *Eng. News Record*, February 1, 1934.
15. Structural Engineers Association of California. *Proc. 44th Annual Conv.*, 1975.
16. Matthiesen, R. B. A Preliminary Survey of Strong Motion Records From the San Fernando Earthquake. Structural Engineers Association of California. *Proc. 40th Annual Convention*, 1971.
17. MacLeod, Iain A. Shear-Wall-Frame Interaction A Design Aid. Portland Cement Association, 1970, Copyrighted.
18. Report of Special Committee on Horizontal Bracing Systems in Buildings Having Masonry or Concrete Walls. Structural Engineers Association of Southern California, 1949.
19. San Fernando, California, Earthquake of February 9, 1971, U.S. Department of Commerce, Vol. III, Geological and Geophysical Studies, p. 139.
20. Repaired R/C Members Under Cyclic Loading (1975), Egor P. Popov and Vitelmo V. Bertero. *Earthquake Engineering and Structural Dynamics* Vol. 4, 129–144.
21. Bond in Interior Joints of Ductile Moment—Resisting R/C Frames Part I: *Experiment* (*1978*), Professors E. P. Popov and V. V. Bertero of the University of California at Berkeley and Grad student S. Viwathanatepa.
22. Applied Technology Council, 480 California Ave., Suite 205, Palo Alto, California 94308

Appendix

Appendix

EARTHQUAKE INTENSITY SCALES

Earthquake intensity scales developed gradually. One of the earliest was the Rossi–Forel Scale that was published in 1883. This scale, which is indicated by the abbreviation R.F., was widely adopted and is still used to some extent, mainly in Europe. The description of earthquake effects in the R.F. Scale does not include modern buildings and is inadequate for present-day needs in other respects. This scale is better suited to European construction and conditions.

A more generally applicable intensity scale was developed by Mercalli in collaboration with Cancani, in 1902. In 1923 Sieberg expanded this scale to embody a wider range of earthquake effects. This scale was used as the basis for the Modified Mercalli Scale of 1931, which is designated by the letters M.M. followed by a Roman numeral indicating the intensity.

The following is a pertinent quotation from the original paper on the M.M. scale, by Woods and Neumann. This paper appeared in the "Bulletin of the Seismological Society of America," Volume 21, No. 4, December 1931.

> The formation of a satisfactory earthquake intensity scale has long been a subject for consideration and discussion among those who are interested actively in studies of shock intensity and its geographic manifestation. The task is more difficult than may appear at first, not only because a large quantity of reliable data is a fundamental necessity, but because questions inevitably arise concerning the application of the

> scale in dealing with phenomena which may be regarded as special cases, or when taking account of special geologic conditions. Most serious, however, is the fact that we do not know exactly what factors combine to constitute intensity as it is ordinarily understood. We are not yet in position to correlate destructive effects with instrumental data so as to establish an adequate measure of intensity. Though the importance of the factor of acceleration is recognized, we have as yet no satisfactory definition of intensity, no formula expressing earthquake violence in terms of ground movement.

Although the above was written in 1931, it is just as true now as it was then. Even now no close correlation has been established between earthquake ground acceleration and building damage. About the best we have is a rough correlation of limited application, such as is stated in the following quotation from a letter to the Author from Dr. George W. Housner, of the California Institute of Technology.

> One of the main points in correlating damage with recorded ground accelerations is the observation that during the San Fernando earthquake no structural damage was sustained by modern code- designed buildings if the peak ground accelerations did not exceed 20%g. Where it approached 30%g there was some structural damage in the form of cracks in beams and columns, etc. Where the ground accelerations approached 40%g severe damage was sustained, but not to the degree that the structures could not be repaired. Where the ground acceleration approached 50%g there was very severe damage to some structures. These observations, it seems to me, should form the basis for setting the levels of design in the building code.

A building vibrates during an earthquake and the response and damage to the structure largely depends on its energy absorption. But this in turn depends on ductility and damping, two indefinite factors. It is likely that evaluation of earthquake intensity will always be determined primarily by observation of physical effects on construction and on natural objects.

In 1956 a new version of the M.M. scale was published by Charles F. Richter. Dr. Richter prefers that this be called the "Modified Mercalli Scale, 1956 Version," rather than the "Richter Scale," which is popularly attached to his magnitude scale. This new version of the M.M. scale follows the original 1931 version very closely, in fact Dr. Richter has simply called it a "restating" of the original. Since however the 1956 version is an improvement in some respects, it is given here rather than the 1931 version.

MODIFIED MERCALLI INTENSITY SCALE, 1956 VERSION

The following comments by Dr. Richter precede the published statement of the intensity scale.

> Some items are omitted for definite reason, and a few additional notes are included, with initials (CFR) to separate them from the scale proper.... Each effect is named at the level of intensity at which it first appears frequently and characteristically. Each effect may be found less stronglyk, or in fewer instances, at the next lower grade of intensity; more strongly or more often at the next higher grade. A few effects are named at two successive levels to indicate a more gradual increase.

Masonry A, B, C, D. To avoid ambiguity of language, the quality of masonry, brick or otherwise, is specific by the following lettering.

Masonry A. Good workmanship, mortar, and design; reinforced, especially laterally, and bound together by using steel, concrete, etc; designed to resist lateral forces.
Masonry B. Good workmanship and mortar; reinforced, but not designed in detail to resist lateral forces.
Masonry C. Ordinary workmanship and mortar; no extreme weaknesses like failing to tie corners, but neither reinforced nor designed against horizontal forces.
Masonry D. Weak materials, such as adobe; poor mortar; low standards of workmanship; weak horizontally.

The following list represents the twelve grades of the scale.

I. Not felt. Marginal and long-period effects of large earth quakes.
II. Felt by persons at rest, on upper floors, or favorably placed.

III. Felt indoors. Hanging objects swing. Vibration like passing of light trucks. Duration estimated. May not be recognized as an earthquake.

IV. Hanging objects swing. Vibration like passing of heavy trucks; or sensation of a jolt like a heavy ball striking the walls. Standing motor cars rock. Windows, dishes, doors rattle. Glasses clink. Crockery clashes. In the upper range of IV wooden walls and frame creak.

V. Felt outdoors; direction estimated. Sleepers wakened. Liquids disturbed, some spilled. Small unstable objects displaced or upset. Doors swing, close, open. Shutters, pictures move. Pendulum clocks stop, start, change rate.

VI. Felt by all. Many frightened and run outdoors. Persons walk unsteadily. Windows, dishes glassware broken. Knicknacks, books, etc., off shelves. Pictures off walls. Furniture moved or overturned. Weak plaster and masonry D cracked. Small bells ring (church school). Trees, bushes shaken (visibly, or heard to rustle—CFR)

VII. Difficult to stand. Noticed by drivers of motor cars. Hanging objects quiver. Furniture broken. Damage to masonry D, including cracks. Weak chimneys broken at roof line. Fall of plaster, loose bricks, stones, tiles, cornices (also unbraced parapets and architectural ornaments—CFR). Some cracks in masonry C. Waves on ponds; water turbid with mud. Small slides and caving in along sand or gravel banks. Large bells ring. Concrete irrigation ditches damaged.

VIII. Steering of motor cars affected. Damaged to masonry C; partial collapse. Some damage to masonry B; none to masonry A. Fall of stucco and some masonry walls. Twisting, fall of chimneys, factory stacks, Monuments, towers, elevated tanks. Frame houses moved on foundation if not bolted down; loose panel walls thrown out. Decayed piling broken off. Branches broken from trees. Changes in flow or temperature of springs and wells. Cracks in wet ground and on steep slopes.

IX. General panic. Masonry D destroyed; masonry C heavily damaged, sometimes with complete collapse; masonry B seriously damaged. (General damage to foundations —CFR). Frame structures, if not bolted, shifted off foundations. Frames racked. Serious damage to reservoirs. Underground pipes broken. Conspicuous cracks in ground. In alluviated areas sand and mud ejected, earthquake fountains, sand craters.

X. Most masonry and frame structures destroyed with their foundations. Some well-built wooden structures and bridges destroyed. Serious damage to dams, dikes, embankments. Large land slides. Water thrown on banks of canals, rivers, lakes, etc. Sand and mud shifted horizontally on beaches and flat lands. Rails bent slightly.

XI. Rails bent greatly. Underground pipelines completely out of service.

XII. Damage nearly total. Large rock masses displaced. Lines of sight and level distorted. Objects thrown in the air.

An improvement of the 1956 version, over the 1931 version, is the better and more definite description of structures and the damage, which is used as a measure of intensity. Also, masonry buildings type C and D are good for this purpose, since they are common in cities everywhere, particulary type D. The following table approximately corresponds the structures in the two versions.

1956 Version	1931 Version
Masonry A	Specially Designed Structures
Masonry B	Buildings of Good Design and Construction
Masonry C	Well Built Ordinary Structures
Masonry D	Poorly Built or Badly Designed Structures

BENDING DEFLECTION OF A FRAME

Nomenclature

L = Distance center to center of end columns in feet.

h = Story height of frame in inches.

A_n = Section area of end columns in nth story.
Transformed area for concrete columns in square inches.

H_n = Distance from nth floor to top of frame in feet.
M_n = Bending (overturning) moment at column inflection points in nth story in pounds feet.
E = Modulus of elasticity of steel for steel frame or concrete for concrete frame.
δ = Lateral deflection of the top of the frame due to direct stress in the columns produced by overturning in inches.

$$\delta = \frac{2h}{L^2 E} \sum \frac{M_n H_n}{A_n} \qquad (19)$$

As already mentioned, the lateral deflection of a frame due to axial stress in the columns from overturning can usually be neglected it is so much smaller than the deflection due to shear. However for an unusually narrow frame (distance center to center between end columns) it may be desirable to include the deflection due to overturning stress in the columns when calculating K_n or the frame stiffness factor S.

The bending deflection of a frame due to axial stress in the columns may be calculated from Equation 19. The summation is for all the floors and stories in the frame. If h varies, include it in the summation as h_n.

Here again, as for shear deflection, the calculated lateral deflection of a concrete frame due to direct stress in the columns will not be as close to the real deflection as would be the case for a structural steel frame.

Index

Index